CRYSTALS AND JOINT DISEASE

CRYSTALS AND JOINT DISEASE

PAUL DIEPPE
University of Bristol
Department of Medicine
Bristol Royal Infirmary
Bristol

PAUL CALVERT
School of Chemistry and Molecular Sciences
University of Sussex
Brighton

LONDON   NEW YORK
CHAPMAN AND HALL

First published 1983
by Chapman and Hall Ltd,
11 New Fetter Lane, London EC4P 4EE
Published in the USA by
Chapman and Hall,
733 Third Avenue, New York, NY 10017
© 1983 P. Dieppe and P. Calvert

Phototypeset by Cotswold Typesetting Ltd, Gloucester
Printed in Great Britain at the
University Press, Cambridge

ISBN 0 412 22150 0

British Library Cataloguing in Publication Data

Dieppe, P. A.
  Crystals and joint disease.
  1. Joints—Diseases
  I. Title
  616.7′2   RC932
  ISBN 0-412-22150-0

Library of Congress Cataloging in Publication Data

Dieppe, Paul.
  Crystals and joint disease.

  Includes bibliographical references and index.
  1. Joints—Diseases. 2. Crystals—Growth.
3. Mineral metabolism—Disorders. I. Calvert, Paul.
II. Title. [DNLM: 1. Crystallization. 2. Joint
diseases—Etiology. WE 300 D562c]
RC932.D53 1983   616.7′2   82-23445
ISBN 0-412-22150-0

CONTENTS

Colour plates appear between pages 166 and 167

v

FOREWORD

This volume has been written jointly by a clinical investigator and a physical scientist. It represents a well-organized, well-written, highly personalized account of the clinical features associated with the three most common crystals found in or near joints. Twenty years ago, the definition of gout changed from one based on clinical description to one based on identification of the characteristic needle-shaped monosodium urate monohydrate (MSU) crystals in joint fluid. Calcium pyrophosphate dihydrate (CPPD) crystals were discovered as a by-product of the development of specific means of identifying MSU crystals. Whatever one's preferred terminology – pseudogout, pyrophosphate arthropathy or CPPD crystal deposition disease – all agree that the definition of the condition is crystallographic. The clinical relevance of the much smaller crystals of the 'apatite' (Gr: deceiver) class is less clearly defined. Here the problem of crystal identification is also much more difficult. Recent work suggests that these deposits may represent a heterogeneous mixture of different calcium phosphate crystals, such as octacalcium phosphate, beta tricalcium phosphate (Whitlockite) and carbonate-apatite in addition to hydroxyapatite itself.

The authors have included excellent succinct descriptions of the structure and function of joints and of the various techniques available for identification of crystals in joint fluids and tissues. In addition to theoretical discussions of nucleation, growth and supersaturation, they have included a practical section on how to identify crystals, how clinically to study patients in whom crystals have been identified, and on treatment of the various articular syndromes associated with crystals.

As pointed out by the authors, this field is evolving rapidly. Not only have the more acidic calcium phosphates mentioned above been identified recently, but evidence for the presence of 'sphere-urates' and of calcium oxalate crystals in joint fluids has been published.

Although no book can be totally current in an evolving subject, this one is unique in the field of the joint crystal deposition diseases. It will be found useful by both investigators and practitioners.

D. J. McCarty
Milwaukee, Wisconsin
1982

PREFACE

Stone formation in the urinary and biliary tracts has been recognized for centuries, and widely accepted to result from inappropriate crystal growth in the body. Many joint diseases, such as gout, are also associated with the presence of crystals, although the discovery of this was only made relatively recently. During the last 20 years there has been a rapid expansion of our knowledge about crystals and joint disease, due in part to the introduction of new techniques for the identification, growth and investigation of crystals.

To date, no book has summarized the necessary understanding of the crystalline state, the technology involved in clinical studies, and the joint diseases now thought to be associated with crystals. This monograph attempts to fill that gap.

In Part One, the theoretical and technical background is covered. The susceptibility of joints to crystal growth and particle–induced damage, and the mechanisms whereby crystals can cause disease are considered. Chapters 3 and 4 provide an introduction to crystal structure: the ways in which crystals form, and the methods of identifying them.

Part Two of the book is clinically oriented, with separate chapters devoted to gout, pyrophosphate deposition, diseases associated with hydroxyapatite crystals, and other particle–related joint conditions. These chapters attempt to explain the diseases by examining why the crystals form, the crystal structure, how these crystals damage the joint, and how these result in the clinical picture described. Present knowledge is insufficient to answer all the questions, and we have also tried to highlight some of the many issues that remain unanswered.

The last two chapters cover treatment, including possible new therapeutic approaches made possible by better understanding of crystals, and summarize our current view of the complex interrelationship between crystals and joint diseases.

The authors are a clinical rheumatologist (PD) and a materials scientist

(PC), and although individual chapters were first written by one or the other of us, they all represent a combined effort. Our aim was to encourage a fully interdisciplinary approach to this area, and for this reason we have tried to concentrate on the main concepts without going into excessive detail on crystallization or pathology. The references are intended to allow the reader to go further in any particular direction without being comprehensive.

The book is intended primarily for practising rheumatologists and research workers involved with joint and stone diseases. It should also be of interest to clinicians and postgraduate medical students in related medical disciplines.

We would like to thank the many friends, teachers and colleagues who have helped with this project. Dr. Barry Shurlock of Chapman and Hall, and the secretarial help of Mrs Margaret Clarke have been invaluable. We would also like to acknowledge the help of Ciba-Geigy in providing a grant for the colour illustrations.

Paul Dieppe, *Bristol, England*
Paul Calvert, *Brighton, England*
1982

Chapter 1

PART ONE

INTRODUCTION

1.1 What is a crystal?

To the early world crystals were unique for their clarity, transparency and reflectivity. Now they can be imitated by polished glass, but the word 'crystal' still conjures up the vision of a clear, shining symmetrical object of great beauty. To the scientist, the crystalline state means that the molecules or ions of a substance are bonded together as closely as possible. This results in the regular and highly symmetrical arrangement that is characteristic of all crystals. Symmetry and tight packing of molecules result in stability and hardness, as well as a high refractive index and related effects on light, which may add sparkle and 'fire' to precious gemstones.

The internal structure of a crystal consists of an ordered repetitive array of atoms which always have exactly the same relationship to their neighbours. This confers their ability to diffract X-rays in the same way as a grating diffracts light. Therefore just as the distance between lines on a re-flecting surface determines the angle of diffraction of a beam of light, so the diffraction of an X-ray beam passed through a crystal is a measure of its internal molecular dimensions. The fact that crystal structure can be precisely determined by X-ray diffraction is fundamental to our modern understanding of chemistry.

Crystals grow by the addition of molecules, one at a time, to the surface. Their external shape therefore reflects the internal symmetry and close packing, and they have facets: flat surfaces with sharp edges. They also break most easily along planes of symmetry, and can be facetted by cleaving, as in some stone cutting. The hardness, facets, and sparkle of many crystals make them precious to industry as well as being symbols of beauty and wealth.

1.2 Crystals and minerals

Crystallography is the study of the form, structure and aggregation of crystals for which a wide range of experimental techniques are used.

1

However, most people who call themselves crystallographers now use X-ray diffraction of crystals to solve molecular structures, and obtain data on lengths and angles of interatomic bonds. Previously the most useful, important and interesting crystals were minerals so that crystallography and mineralogy grew up together. They have separated since the synthetic chemist and biochemist provided more to interest the crystallographer.

Strictly, a mineral is a natural compound produced by inorganic processes, but this leaves in doubt the status of such biological products as coal and amber and even the carbonate crystals in chalk and limestone. Since minerals are often mixed together in rocks, optical methods have continued to be important for identification in mineralogy. Many minerals are not crystalline but amorphous; glass on the lunar surface is an example of current interest.

Within these terms it seems incorrect to call a crystalline deposit which forms in bone or in a joint a mineral, even though identical crystals may occur as minerals in rock. However, if crystallographers can become mineralogists at will, it seems unfair not to allow crystals the same privilege. Thus we will use mineral in the biological context to mean precipitated, inorganic materials such as calcium phosphates or carbonates.

1.3   Crystals in biology

Many living things form crystals. There are the crystalline fibrous polymers such as cellulose, keratin and collagen which have complex structures that put them outside our simple ideas of single crystals. Other crystals have specific uses, such as the oriented uric acid needles in the shell of the Scarab beetle which enhance its iridescent colour, or the cutting hardness of hydroxyapatite crystals in tooth enamel. However, the main purpose of biological crystals is to provide a stable rigid skeleton to support the organism. This can be a largely mineral aggregate, like coral or mollusc shells, or a mineral reinforcement of an organic matrix as in bone or crustacean shell. The mineral makes the skeleton much more rigid and, especially in the case of sea shells, costs the animal little in energy or resources to produce. The disadvantage is that the weight is increased. Thus weight is probably important to insects which have an organic shell, whereas curstaceans, which spend much of their time in water, have mineralized shells. Since phosphorus is not plentiful we must conclude that slender mineral-reinforced bones are preferable to a thicker wholly organic skeleton of the same stiffness.

1.4 Crystal deposition diseases

A crystal deposition disease may be defined as a pathological condition associated with the presence of crystals which then contribute to the tissue damage.

This carefully worded statement implies that crystals can cause harm as well as benefit to living organisms. It also suggests that the presence of crystals in an area of damaged tissue can result from disease, rather than cause it. Plants may 'mineralize' on death (as in the formation of the petrified forests of Arizona), and dead or damaged animal tissue often calcifies. But crystals can also cause damage, and synovial joints are peculiarly susceptible to both deposition of crystals and crystal-induced damage.

A central theme of this book is the belief that many pathological processes are associated with the formation of crystals, which can then damage the tissues in a variety of ways to influence the resulting disease. Crystal deposition is thus seen as one of a number of processes which together alter the anatomy or physiology of complex organs such as synovial joints.

1.5 Historical background

One of the earliest descriptions of a crystal deposition disease appears in the Bible

'But his wife looked back from behind him and she became a pillar of salt' (*Genesis* 19, 26).

However, the crystal deposition disease best known to ancient medical practitioners was stone formation. Bladder calculi have been recovered from Egyptian mummies 7000 years old, and ancient Greek medicine recognized urinary stone formation: Hippocrates commenting 'I will not cut, even for the stone, but leave such procedures to the practitioners of the craft'. Surgery has not been short of eager practitioners; a variety of instruments have been available through the centuries for performing lithotomy (removing the stone, or crystalline deposit, from the bladder). Galen treated stone disease with wine and honey, parsley and caraway seed, but other famous men, including Celsius, preferred lithotomy, and the latter's operation was used from the 1st to the 17th Century. Such is the misery of this form of crystal formation that some have even operated on themselves, such as the Amsterdam Blacksmith Jan de Doot, who removed his own bladder stone with a kitchen knife in 1651!

The idea that joint diseases might be caused by crystal deposition is relatively recent. Gout has a long and famous history (see Chapter 6), and has been a well-described clinical entity for centuries. The Roman physician

Arelaeus was one of the first people to suggest that a specific toxic substance might be the cause of gout, but the first to suggest an involvement of crystals was Sir Alfred Baring Garrod (1819–1907) (Fig. 1.1). This theory was not

Fig. 1.1 Sir Alfred Garrod (1819–1907). From an original photograph in the Wellcome Historical Medical Museum and Library.

widely accepted until the early 1960s. At that time polarized light microscopy was used to examine joint fluid, and it was shown that all cases of acute gout, and only gout, were associated with the presence of urate crystals. Some earlier work of the German physicians Freudweilher and Hiss was rediscovered, showing that urate crystals could cause an inflammatory reaction like that of clinical gout, and so the jig-saw pieces seemed to fit. A high serum urate level causes crystallization, and the crystals cause the inflammation of gout. As explained in Chapter 6, the story of gout no longer seems quite as simple as that, although the central role of crystal deposition is certain.

Joint radiographs, as well as light microscopy, led to the discovery of calcification of joint cartilage: the formation of calcium phosphate crystals. The first calcium salt to be implicated in joint disease was calcium pyrophosphate dihydrate, although more recently, analytical electron microscopy and other sophisticated techniques have led to the identification of other calcium phosphates in pathological joint tissue, including hydroxyapatite (bone mineral).

1.6 Crystal deposition diseases of joints

The three principal crystalline deposits described in association with joint disease are urate, pyrophosphate and hydroxyapatite (Table 1.1). Urates are associated with gout, which principally affects the peripheral joints, often starting in the foot or hand. Pyrophosphate deposition occurs most commonly in the knees, wrists and pelvis. Hydroxyapatite is often deposited in periarticular tissues, especially around the shoulder and sometimes in the hip, spine and peripheral joints as well. The three deposits, therefore, differ

Table 1.1 Crystals and other particles associated with joint diseases

(a) *The three major crystal deposition diseases of joints*

Crystal	Associated diseases	Distribution in joints
1. Monosodium urate monohydrate	Acute and chronic gouty arthritis	Mainly *peripheral* (hands and feet)
2. Calcium pyrophosphate dihydrate	Acute pseudogout	*Intermediate* (knees, wrists and pelvis)
	Chronic pyrophosphate arthropathy	
3. Hydroxyapatite	Acute calcific periarthritis	Mainly *central* (shoulders, hips spine)
	Chronic destructive joint disease (osteoarthritis?)	

(b) *Other crystals found in joints*
1. Dicalcium phosphate dihydrate and other calcium phosphates
2. Crystalline steroids
3. Cholesterol and calcium oxalate

(c) *Other particles that may be found in joint tissue*
1. Bone and cartilage fragments
2. Fragments of metal prostheses or cement
3. Thorns, foreign bodies and other particles penetrating through skin

in their distribution as well as in their characteristics and metabolic background (Table 1.1). Each is associated with both an acute inflammatory reaction and a chronic destructive form of joint disease, and between them these conditions make an important contribution to the crippling rheumatic diseases.

Other crystals are sometimes deposited in joints, and may contribute to damage. Crystals may also be injected into the joint cavity from outside, as in the therapeutic use of crystalline steroid preparations. Other particles, such as fragments of cartilage or bone, or plant thorns penetrating the skin, may also cause joint disease. Thus the crystal deposition diseases could be considered as one form of a wider group of particle–induced conditions.

These conditions are described in later chapters of this book. First, the synovial joints and their susceptibility to crystal deposition (Chapter 2), and the structure and origin of the crystals (Chapter 3) are discussed. These chapters provide the background and framework of our approach to the crystal–induced arthritis.

Further reading

BUNN, C.W. (1964) *Crystals: Their role in nature and science*, Academic Press paperback, New York and London.

HURLBUT, C.S. and KLEIN, C. (1977) *Manual of Mineralogy* (after James Dwight Dana), 19th edn, Wiley, New York.

Chapter 2

JOINT FUNCTION
AND EFFECTS OF CRYSTALS

2.1 Introduction

A joint allows relative motion between the parts of a rigid skeleton. In principle, six degrees of freedom are possible, sliding in three directions and rotation about three axes. In practice, the requirement that the joint does not fall apart under load usually limits motion to fewer directions. For instance, the human humero–ulnar joint at the elbow is essentially a hinge joint, only allowing rotation about one axis, with the surfaces so shaped as to prevent sliding parallel to that axis. The hip and shoulder are ball and socket joints, allowing rotation about three axes. A system of ligaments across these joints limits the motion by becoming taut at different positions. The tempero–mandibular joint (jaw) does have the full six degrees of freedom by being a loosely fitted joint surrounded by a bag of muscle which helps to prevent dislocation.

In arthropods joints are often very simple. A crab's leg joint consists of a double pin and socket connection between two sections of hard cuticle with a flexible section of cuticle allowing them to move relative to one another. This allows motion about one axis and is supplemented by a second neighbouring joint with a second perpendicular axis of rotation.

In vertebrates the compressive forces acting on the internal skeleton makes low–friction sliding joints a necessity if movement is not to consume large amounts of energy. Hence the synovial joint, in which the bones are capped by hyaline cartilage and lubricated by synovial fluid, has a friction coefficient of about 0.005, comparable to that of oil–fed engineering bearings. (The friction coefficient of an ice skate on ice is about 0·02.) This design also requires a large bearing surface area to avoid overloading the cartilage.

Normal joint surfaces are not congruous; in other words only a small section of the surface is load–bearing at any one time, or in any one position.

Joint congruity does increase with age and in certain diseases, and by altering the forces acting on the surface may contribute to joint wear. In addition, the vertebrate low–friction joint construction requires systems to resist tensile forces which tend to cause disarticulation. Ligaments tighten up to restrict certain movements, but muscles too have an important part to play in maintaining stability. The tendon–bone insertion (enthesis) is richly innervated and may play a vital role in signalling to the central nervous system the forces applied to a joint, allowing feedback control of muscle tone and therefore joint control. The sliding motion of the opposed bones also makes synovial joints vulnerable to wear by hard, loose particles such as crystals (see Chapter 5).

2.2   Classification of joints in man

There are three main types of joint (Table 2.1): (1) the rigid non-moving synarthrosis, (2) the amphiarthrosis, which is capable of slight movement,

Table 2.1   Classification of joints

Functional	Structural	Examples
(1) Synarthrosis (Non-moving)	Fibrous-syndesmosis	Skull
(2) Amphiarthrosis (Slight movement)	Cartilagenous- synchondroses and symphyses	Inter-vertebral joints
(3) Diarthrosis (Free moving)	Synovial, 'cavity' joints	Limb joints

and (3) the diarthrosis which is a free moving articulation. Synarthroses are held together by dense fibrous tissue (syndesmoses), and the best examples are the joints between the bones of the cranium. Amphiarthroses are cartilagenous joints (synchondroses and symphyses), in which slight movement due to compressibility of the cartilage is possible. Examples include the intervertebral joints of the spine and the pubic symphysis. The diarthroses are free moving synovial joints, and are the only ones which contain a cavity between the bone ends. They are the most important with respect to crystal deposition diseases, and form a majority of the articulations of the limbs.

2.3   Development

The joints develop between the fourth and eighth weeks of foetal life. They arise within the primitive limb bud mesenchyme which early on in limb development starts to show signs of clefts which later develop into the cavities of synovial joints. The mesenchymal cells condense at the margin of these clefts and later form the developing synovium and capsule surrounding the cavity. The area lining the cavity is the chondrogenic zone, which develops into the hyaline articular cartilage. Relatively little is known about factors that control this very early differentiation, although, as mentioned later, there are a number of factors which are important in maintaining the integrity and specialized nature of the involved tissues.

2.4   Synovial joint structure

A stylized cross–section of a synovial joint is shown in Fig. 2.1. The bone ends are covered by hyaline articular cartilage which forms the sliding surfaces of the joint. The small cavity is lined by a thin vascular membrane, the synovial membrane, which rests on sub-synovial connective tissue that merges into the joint capsule encasing the whole structure. The capsule is

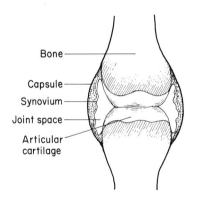

Fig. 2.1   Schematic cross-section of a synovial joint.

thickened in various areas to form the restraining ligaments of the joint, which hold the parts together. In some joints (Table 2.2), extra stability and mechanical control is achieved by the presence of fibrocartilage pads within

Table 2.2   Joints containing fibro–cartilage pads (Menisci)

Knee
Tempero–mandibular
Sterno–clavicular
Distal radio–ulnar
Acromio–clavicular (inconstant)

the joint space. An example is the knee joint where the convex condyles of the lower end of the femur roll on the flat upper end of the tibia. A fibrocartilage wedge surrounding the contact areas stabilizes the joint against slippage and probably distributes much of the load. These menisci are particularly important in pyrophosphate arthropathy as the crystals involved deposit preferentially in fibrocartilage (see Chapter 7). The joint cavity contains synovial fluid which derives its name from its egg white gelatinous quality. (The word 'synovial' was first used by Paracelsus in the 16th Century.)

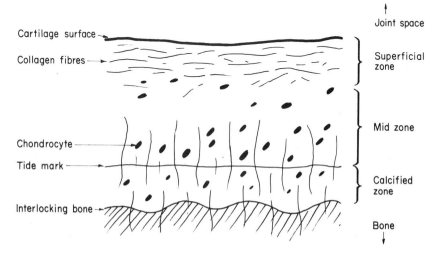

Fig. 2.2 Diagrammatic representation of the structure of hyaline articular cartilage. Note: *Superficial zone:* parallel horizontal arrangement of collagen fibres; suggesting flattened cells. *Mid zone:* vertical collagen fibres; rounded chondrocytes. *Calcified zone:* partially mineralized (hydroxyapatite).

Some aspects of the structure of *articular cartilage* are shown in Fig. 2.2. This white substance forms a 2 to 4 mm thick covering over the bone ends of all synovial joints. It is avascular, alymphatic and is not innervated. As shown in the figure, the bone end is undulating and interlocks with the lowermost calcified zone of the articular cartilage. This is the only zone into which blood vessels penetrate from the bone. The remaining middle and upper zones of the cartilage contain a small number of cells (chondrocytes) whose shape varies according to their position and which often exist in clumps. They are mainly found in the lower zone. The bulk of the structure is made up of *intercellular ground substance*, a swollen gel containing about 75% water, 15% (60% dry weight) collagen fibres and 5% (20% dry weight) proteoglycans.

The basic structural unit of collagen is three polypeptide chains wound into a rod-like triple helix of about 1·4 nm diameter by 300 nm long. The helices are packed into fibrils and the fibrils into fibres which are about 30–80 nm in diameter and many microns long. In most mammalian collagen (type I) two of the chains in the helix are identical (α_1 I) while the third (α_2) has the same length but a slightly different amino acid composition. In cartilage the collagen (type II) contains three identical chains ((α_1 II)$_3$) but the significance of this distinction is unknown. The fibres are linked in some way to form a strong open mesh.

At the outer cartilage surface the collagen fibres are closely packed and oriented parallel to the surface; in each part of the surface they also lie predominantly parallel to each other so that the surface layer will have locally strong and weak directions. Below the surface the fibre orientation is random, but dependent on the state of stress of the cartilage. At its base articular cartilage is calcified and the fibres lie perpendicular to the bone surface. This arrangement and the wavy cartilage–bone interface ensure a strong bond between the two structures. The chondrocytes are surrounded by a distinct finely textured zone of pericellular matrix up to 3 μm thick which is rich in proteoglycans.

As shown in Fig. 2.3 the proteoglycans of the matrix consist of glycosaminoglycan chains predominantly of chondroitin–6–sulphate and keratan sulphate arranged like a bottle-brush about a protein core. These complexes of several millions in molecular weight seem to be bound to hyaluronic acid molecules by a special binding protein. The keratan sulphate chains may bind to collagen fibres, and this and other possible mechanisms binding proteoglycan to collagen help to form a cross–linked gel within the collagen meshwork. The composition of the proteoglycans changes with the site and age of the cartilage. Thus mature intervertebral disc fibrous cartilage has a greater ratio of keratan sulphate to chondroitin sulphate. However, the relative chondroitin sulphate content of all cartilage also decreases markedly with age, from about 70% of the glycosamino-glycans at $4\frac{1}{2}$ to 40% at 76 years of age (Gower and Pedrini, 1969). The carboxylate and sulphate groups on the chondroitin are ionized at physiological pH, making the chains strongly anionic. These negative charges together with their complementary sodium ions create a high local charge density within the gel. This results in an osmotic pressure so that water swells the gel. However, this swelling is opposed by the collagen network so the cartilage becomes rigid after the manner of a balloon inflated inside a string bag. This structure also means that when the cartilage is compressed slowly, fluid will be squeezed out, whereas under rapid loading, where there is no time for fluid flow, the material will appear rigid.

Fibrocartilage pads have a rather different structure and some compara-tive aspects of hyaline and fibrocartilage are outlined in Table 2.3. They

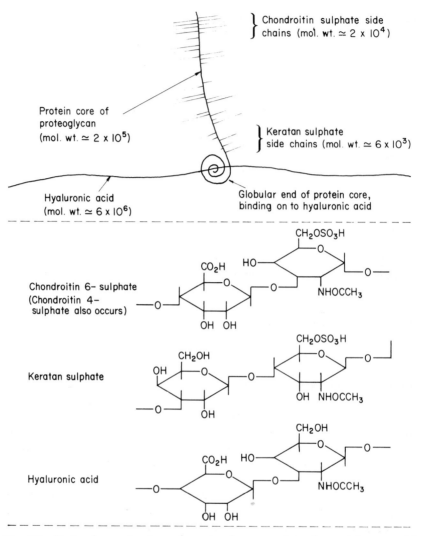

Fig. 2.3   Proteoglycan structure. A proteoglycan consists of a protein core with attached side chains of the sugars chondroitin and keratan sulphate. These are bound to the high molecular weight polysaccharide, hyaluronic acid, resulting in massive complexes.

contain a similar amount of water to hyaline articular cartilage, but the solid contains more collagen and has considerably less proteoglycan ground substance, implying a less rigid, more elastic, material. The collagen is of type I, and is associated with some elastic tissue. As in the case of hyaline cartilage the total amount of solid cannot be fully accounted for by collagen,

Table 2.3 Chemical composition of various forms of cartilage in two species. Non-collagenous proteins, lipids and other materials form the percentage of dry weight not accounted for by collagen and glycosaminoglycans

| Type of cartilage | Water content (%) | % dry weight | | | Cell density (cell mm³ tissue) |
		Collagen	Glycosamino-glycan	Elastin	
Human					
Epiphysial	81	37	15	—	—
Bronchial		37	11	—	—
Costal	61	38	6	—	11 000
Articular, femoral head	72	66	18	—	10 000
Fibrocartilage, meniscus of knee joint	74	78	2·4	0·6	12 000
Bovine					
Epiphysial	76	51	22	0	—
Nasal	74	35	43	0	—
Articular	78	72	14	0	20 000
Fibrocartilage, meniscus	66	82	2	0	—
Elastic, auricular	71	53	12	19	—

Reproduced with permission from Stockwell (1979).

elastin, and proteoglycans. There are small amounts of glycoprotein, and other as yet unidentified macromolecules, within these structures.

The *synovial membrane* is another specialized tissue crucial to the structure and function of synovial joints. It is a vascular membrane which has an irregular cell lining, normally only one to two cells thick. These cells sit on a loose matrix of connective tissue. Although highly vascularized, it is not a true membrane as there is no continuous basement membrane, and clear gaps can be seen between the cells in the capillary loops. As shown in Fig. 2.4, there are three different areas of synovial membrane. The main part of it is a folded structure which can expand or concertina according to the movement of the joint. There is then a 'synovial sink area', adjacent to the third part of the membrane, which is attached to the junction between articular cartilage and bone. There are two chief types of cell in the synovial lining; about one third of them are monocytic in type, and they are able to phagocytose foreign material to clear debris (including crystals) from the joint. These are the Type A cells. The remaining two thirds, the B cells, are more like fibroblasts or chondrocytes in appearance, and are synthetic being responsible for the production of the hyaluronate present in the synovial fluid.

Some years ago it was thought that synovial fluid, the viscous substance

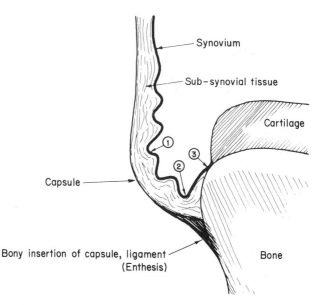

Fig. 2.4 Synovial joint structure. Types of synovium: (1) folded membrane expanding and contracting with movement, (2) synovial 'sink' area, (3) attachment to bone–cartilage junction.

present in the space of all joints in quantities of 1–2 ml, consisted of a dialysate of blood plus secreted hyaluronic acid, which is a glycosamino-glycan of molecular weight about 2×10^6 with some associated protein. There are, however, some other differences in the concentration of various substances between the synovial fluid and the blood. This is partly due to the charge effect of the anionic glycosaminoglycans (the Donnan equilibrium), but it is becoming clear that the synovium is a complex 'blood–joint barrier'. The main functions of the fluid are the nutrition of the avascular cartilage, and lubrication of the joint. In the normal joint there are few cells present in synovial fluid, the average is probably in the order of 200–500 cells per cubic millimetre, most of which are either derived from the synovial membrane or are blood-derived polymorphonuclear leukocytes.

The whole joint is enclosed by a capsule which contains parallel bundles of collagen and a few fibroblast-like cells. This becomes thickened with large aggregations of collagen to maintain stability according to the various directions of movement in a particular joint. The area of attachment of the capsule to the periostium and bone is a specialized structure similar to the enthesis described in Chapter 8, and is susceptible to the formation of insoluble calcium phosphates (Fig. 2.4).

2.5 Function

2.5.1 Lubrication

Friction coefficients define the ratio of the force necessary to start one body moving over another to the load pressing them together. Typical values are 1 for a tyre on a dry road, 0.05 for a plain, oiled bearing and 0.001 for a ball bearing. Intact joints show values of 0·003–0·015 if steps are taken to remove the forces due to the capsule and ligaments. A cartilage-covered bone end rubbing on glass in fluid gives values of 0·001–0·07. These values are sufficiently low to imply that some fairly effective mechanism of lubrication is operating, but it is not clear what. Replacement of the synovial fluid with saline increases the friction but not dramatically. The friction does increase with time up to about 30 min and decreases slowly if the joint is unloaded, suggesting that the cartilage is deforming to increase the degree of contact. Swanson discusses various lubrication mechanisms and favours boundary lubrication due to adsorption of a layer of hyaluronic acid and protein on the cartilage surface; certainly the viscosity of the synovial fluid itself does not seem to be important but the presence of hyaluronic acid protein complex does. Despite one's intuitive feelings to the contrary, there is no evidence that fibrillation of cartilage increases the joint stiffness or that a failure of lubrication due to, for instance, rheumatoid synovial fluid causes fibrillation.

2.5.2 Load carrying and fracture

The purpose of articular cartilage is apparently to spread the loading between bones so that high local stresses at contact points do not lead to bone fracture. Although it would make a good damping material it probably has little effect in opposing shock loadings because it is a relatively thin layer; damping in the bone itself will therefore be more important. The tensile stiffness and strength of cartilage are really determined by the collagen network, but the resistance to compression and indentation is mainly a function of the proteoglycan content. On loading cartilage shows an initial elastic response, followed by a slow creep to a new equilibrium over about 30 min as water is squeezed out of the gel (Fig. 2.5).

Cartilage can be fractured by high loads but is also sensitive to fatigue, repeated application of loads below the fracture level finally causing damage to occur. In this respect cartilage resembles other materials where fibres reinforce a relatively soft matrix such as glass fibre-reinforced polyester (fibreglass). Damaged or degrading cartilage softens and fibrillates where loss of proteoglycan apparently gives rise to a loss of bonding between the collagen fibres. The chain of cause and effect accompanying fibrillation is

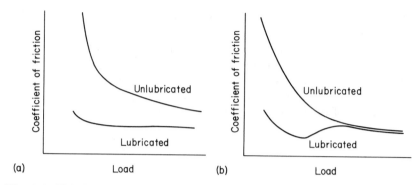

Fig. 2.5 Friction in joints. The difference in the coefficient of friction of lubricated and unlubricated joints on sudden or static loading. (a) The reduction in coefficient of friction on increase in loading suggests that fluid may be expelled from the cartilage to aid lubrication ('squeeze-film' lubrication). (b) This advantage is lost with prolonged high loads. After Unsworth *et al.* (1975).

unclear, but a probable sequence is that some general weakening of the cartilage or abnormal loading is followed by fracture of the collagen net. This leads to loss of proteoglycan and hence to fibrillation which spreads through the cartilage as the proteoglycan progressively leaks out through the fissures. An alteration in chondrocyte function leading to either altered proteoglycan content or increased catabolic enzyme production could be primary events increasing the susceptibility of the collagen network to load.

An alteration in the mineral (crystal) content of subchondral bone could alter its load-damping ability and alter the load on cartilage. There is some evidence that loss of mineral (osteopaenia) does protect against osteoarthritis supporting this hypothesis. Conversely, the presence of crystal deposits in the cartilage itself could alter its compliance and also predispose to cartilage damage in response to loading. Once an area of fibrillation develops the contact area and congruity of the joint may alter, leading to a sequence of slow, progressive damage to cartilage.

2.5.3 Nutrition

Since cartilage is avascular the chondrocytes must derive their nutrition either via the bone or the synovial fluid. Measurements have shown that in mature joints the subchondral bone plate prevents significant transport of nutrients from the underlying bone. The supply is almost totally by diffusion in from the synovial fluid whence it arrives from the highly vascular synovium. The gel structure of the matrix prevents most large molecules from penetrating the cartilage but permits the flow of small molecules and ions.

Studies by Maroudas (1980) and Hadler (1981) suggest that transport of some small ions such as calcium is facilitated by cartilage proteoglycans. The ability of the anionic saccharide chains to bind calcium and other cations may also influence the availability of free ions, and thus the susceptibility of the tissue to deposit calcium salts.

2.6 Ageing of connective tissue

Ageing and death are compulsory. In the first two decades of our lives we are growing, and connective tissues are developing into the final form of the skeleton, joints and associated tissues. Later, as we get older, bones get thinner, we become shorter in stature, the skin texture changes, and so do the joints. Crystal deposition diseases appear mainly in later years. It is relevant, therefore, to consider what changes are occurring in joints as age advances.

Joints develop from the pluripotential mesenchymal cells and the fibroblasts, synovial lining cells, osteoblasts and chondroblasts are all of similar type, showing specialist function according to their site. What factors control this are unclear, although feedback control dependent on physical stresses in the surrounding matrix and chemical changes in surrounding fibres and ground substances are obviously important. For example, cells seem to behave differently when in a culture dish, than when they are in their normal environment in the joint. Recently an increasingly large number of intercellular transmitter substances have been defined, which when secreted from one cell alter the behaviour of another. Macrophage-derived factors which alter synovial cell function have been found and synoviocytes in culture produce substances which induce chondrocytes to 'auto-digest' cartilage in culture. Cell numbers and activity alter with age, and a change in balance of these inter-cellular transmitters could explain some aspects of ageing.

There is relatively little change in the properties or composition of articular cartilage with age. The collagen is essentially inert and shows very little turnover or change with ageing. Proteoglycans do slowly turnover and are continually produced by the chondrocytes, but little is known about the process or their half life. The proteoglycan content of cartilage decreases slightly with age and the deep layers show an increase in keratan sulphate at the expense of chondroitin sulphate. Articular cartilage also takes on a yellow colour with ageing, and becomes a little thinner.

Whilst these changes are minor, there is also an increase in the number and severity of fibrillated regions with age. Since these regions of cartilage destruction are localized they cannot be regarded as a direct consequence of ageing. Fibrillation is widely believed to be the primary lesion of osteoarthrosis although it may remain localized and not progress to the large-scale destruction seen in this condition. The extent to which age associated changes and fibrillation contribute to osteoarthritis is disputed; Gardner (1978), for example, arguing that age changes are important; whereas Sokoloff (1978) and others take a different view. The demonstration by Byers *et al.* (1970) that there are quite different types of fibrillation in different areas of the femoral head, the progressive and non-progressive lesions, suggests that age-induced changes must interact with some other factors before leading to osteoarthritis.

2.7 The deposition of crystals in joints

Synovial joints are especially prone to crystal deposition diseases. This is well illustrated by the example of gout (Chapter 6). Soluble uric acid and sodium ions are both distributed uniformly throughout all body compartments, but the relatively insoluble salt, monosodium urate monohydrate, forms only in the connective tissues, and preferentially in joint cartilage. Furthermore, deposits of urate in the joints are usually associated with symptomatic arthritis, whereas deposits at other connective tissue sites (such as the cartilage of the ear lobes) may be asymptomatic and go unnoticed for many years. There are therefore two aspects to the susceptibility of joints to gout and other crystal deposition diseases. First the tendency for crystals to form in synovial joints, and secondly their ability to cause clinically overt problems.

Cartilage is a natural precursor of bone, and contains all the elements necessary for the formation of the collagen matrix and hydroxyapatite mineral that make up bone. As outlined in Chapter 8, osteoarthritic hyaline cartilage shows both metabolic and morphological changes which produce a tissue that is similar to growth cartilage. For example, the chondrocytes become metabolically more active, the levels of alkaline phosphatase increase, and there is a change in the structure of the proteoglycans. The chondrocytes also form matrix vesicles which may mineralize; these are all features of actively growing cartilage from which bones form. Thus in osteoarthritis at least, the normal resting hyaline joint cartilage can undergo a metaplastic type of change. The process is analogous to synovial osteochondromatosis, in which a metaplasia in cells of the synovium or capsule, allows formation of cartilage rests, and later calcification and formation of true bone matrix. Metaplastic change is therefore one mechanism by which crystal deposition can occur in joints, and explains in part the tendency to form bone mineral in a number of pathological joint conditions.

The hormonal, chemical and physical factors which control cell function in connective tissues are not fully understood. However, normal joint cartilage has a number of special factors that tend to inhibit the formation of bone mineral, and perhaps of other crystal deposits (Table 2.4).

Table 2.4 Possible factors inhibiting crystal deposition in joint cartilage

Avascularity
Proteoglycan aggregates ⎫
Glycoproteins   ⎬ Macromolecular
Hyaluronate   ⎭ crystallization inhibitors
Inorganic pyrophosphate

The avascularity of articular cartilage is a significant factor. Remodelling of bone and the increased bone formation occurring in osteoarthritis are both dependent on the influx of new blood vessels in the subchondral bone area and the calcified bone cartilage. A special factor promoting this revascularization of cartilage has been implicated in joint disease such as osteoarthritis, and it appears there may also be inhibitory substances that normally prevent the ingrowth of vascular tissue. As bone formation and a good blood supply seem to go closely hand in hand, this may well be a major factor in the inhibition of crystal formation.

The nature of the ground substance of cartilage may also be inhibitory. Many macromolecules, charged and uncharged, are known to be inhibitors of crystal growth in various systems, presumably because of their propensity for binding to surfaces. Proteoglycan, glycoprotein or hyaluronate may play this role.

Inorganic pyrophosphate $P_2O_7^{4-}$ has also been suggested as a local controlling factor inhibiting the formation of apatite. $P_2O_7^{4-}$ ions can inhibit hydroxyapatite growth, and these are normally formed by the active chondrocyte, although whether appreciable quantities are transported into the extracellular medium without cell damage seems unlikely.

A number of factors special to connective tissues of joints, which may also help to explain their enhanced susceptibility to depositions, are tabulated in Table 2.5. As well as being avascular, cartilage is relatively acellular. One consequence of this is that there is a paucity of scavenger cells such as macrophages and polymorphs present to remove any particles that may form within the matrix of the tissue. Thus the formation of small crystals of urate may be common throughout the body in gout. However, in other tissues these will be removed by scavenger cells.

Table 2.5 Possible factors favouring crystal deposition in cartilage

---

Acellularity
Mechanical forces
Transport properties of cartilage
Collagen fibres
The physical matrix
Loss of normal inhibitory mechanisms (Table 2.4)

Another factor may be the mechanical loads to which cartilage is subjected. The resultant flexing of the cartilage could fracture crystals within it. The fragments will serve as new nuclei and may initially grow at enhanced rates due to their irregular surfaces. Thus the net crystallization rate will be much faster.

The limited transport properties of cartilage alluded to are also likely to be important. In the case of pyrophosphate arthropathy the crystal tends to form in a line in the mid-zone of the cartilage. This is the region where a maximum concentration product would occur if pyrophosphate were produced uniformly in the cartilage and diffuses out to the synovial fluid while calcium diffuses in from the fluid. Such bands of precipitation are seen in experimental precipitations in gels where two reagents travel towards one another. The characteristic line shadow shown in the radiographs in Chapter 7 is very suggestive that this mechanism can operate. The collagen fibres have been investigated extensively as possible sites of nucleation of hydroxyapatite crystals, as mentioned in the next chapter.

Finally, breakdown of any of the inhibitory mechanisms outlined in Table 2.4 may help to promote the formation of crystals. The tendency for joint bone ends to remodel with age, and for the proteoglycan ground substance and metabolic activity for the chondrocytes to alter has already been mentioned, and may well promote the critical breakdown in inhibition, and explain the increasing prevalence of crystal deposition diseases with increasing age. Notwithstanding the possible role of the mechanisms outlined above, the extreme specificity with which crystal deposition diseases do affect cartilage is remarkable. Many tissues of the body contain similar components, that is collagen and proteoglycans, and a number of other tissues are avascular. The fact that cartilage is singled out, and that specific parts of specific joints are usually affected first is even more surprising. Part of the solution to this must be that crystals are not usually sought in asymptomatic parts of the body and a careful survey for signs of limited deposition elsewhere in gout, for instance, would be of great interest.

Nonetheless, it does seem that articular cartilage is especially vulnerable to crystal deposition.

2.8   Crystal-induced damage to joints

A crystal deposit within or around a synovial joint is often associated with disease of the tissue, and, as already mentioned, the joints seem particularly susceptible to the consequences of deposition as well as to the formation of the crystal.

Crystal-induced damage is considered in detail in Chapter 4. In this section three concepts presented in Table 2.6 will be briefly introduced. First, the relative inability of the cartilage to remove particles that may form within its matrix has already been mentioned. In contrast to this the synovial membrane with its large number of phagocytic monocyte-type cells would appear to have foreign body removal from the joint space as a main function, and it could be argued that removal of particles from the joint fluid is therefore important to prevent joint damage. Many of the crystal-related diseases follow a course starting with crystal deposition in a cartilage, followed by a shedding of those crystals into the joint space, and with subsequent removal by cells from the synovial membrane.

Table 2.6 The susceptibility of synovial joints to crystal-induced damage

(1) Impaired removal of particles
(2) Mechanical factors
(3) Synovial membrane inflammation

Mechanical factors are also of obvious importance. Crystal deposits may weaken cartilage, resulting in enhanced joint destruction in the presence of crystal deposits. Cell lysis by phagocytosed crystals may also be exacerbated by the forces on the cell between the bearing surfaces of the joint.

The synovial membrane is also peculiarly susceptible to inflammation, and many other joint diseases are dominated by a relentless inflammatory process occurring in the synovium, with apparently little stimulus to keep it going. Inflammation is a major feature of many crystal deposition diseases and appears to be induced by crystals in many instances.

2.9   Summary

Joints consist of highly specialized tissues, continually subjected to high mechanical stresses. As age advances, joints become more and more susceptible to the formation of crystals, either due to breakdown of normal inhibitory mechanisms, or due to intrinsic factors that may explain why insoluble products tend to form first in the connective tissues. The mechanical forces present in the joint, as well as its vascular synovial lining, appear to make it particularly susceptible to damaging consequences when the crystals form. Thus the synovial joint is a prime target for crystal deposition diseases.

Further reading

COMPER, W.D. and LAURENT, T.C. (1978) Physiological function of connective tissue polysaccharides. *Physiol. Rev.* **58**, 255.

FREEMAN, M.A.R. (1973) *Adult Articular Cartilage*, Pitman Medical, London.

HALL, D.A. (1976) *The Aging of Connective Tissue*, Academic Press, London.

HASSELBACHER, P. (1981) The biology of the joint. *Clinics in Rheum. Dis.* 7(1).

NUKI, G. (1980) *The aetiopathogenesis of Osteoarthritis*, Pitman Medical, London.

STOCKWELL, R.A. (1979) *The Biology of Cartilage Cells*, Cambridge University Press, Cambridge.

VAUGHAN, J.M. (1975) *The Physiology of Bone*, 2nd edn, Clarendon, Oxford.

VINCENT, J.F.V. and CURREY, J.D. (eds) (1980) *The mechanical properties of biological materials*, Soc. Exp. Biol. Symp. 34, Cambridge University Press, Cambridge.

WAINWRIGHT, S.A., BIGGS, W.D., CURREY, J.D. and GOSLINE, J.M. (1976) *Mechanical design in organisms*, Edward Arnold, London.

Text references

BYERS, P.H., CONTEPOMIC, C.A. and FARKAS, T.A. (1970) Post-mortem study of the hip joint. *Ann. Rheum. Dis.* **29**, 15.

GARDNER, D.L. (1978) Diseases of Connective Tissue, *J. Clin. Path.* **12**, (suppl.).

GOWER, W.E. and PEDRINI, J. (1969) Age related variation in proteinpolysaccharides. *J. Bone Joint Surg.*, **51A**, 1154.

HADLER, N. (1981) *Clinics in Rheum. Dis.*, 7, 71.

MAROUDAS, A. (1980) in *Studies in Joint Disease I* (eds A. Maroudas and E.J. Holborow), Pitman Medical, London.

SOKOLOFF, L. (1978) *The Joints and Synovial Fluid*, Vols. I and II, Academic Press, London.

STOCKWELL, R.A. (1979) *Biology of Cartilage Cells*, Cambridge University Press, Cambridge.

UNSWORTH, A., DOWSON, D. and WRIGHT, V. (1975) Some new evidence on human joint lubrication. *Ann. Rheum. Dis.*, **34**, 277.

Chapter 3

CRYSTALS AND THEIR DEPOSITION IN JOINTS

3.1 Crystals

3.1.1 The crystalline state

There are three principal states of matter: crystal, liquid and gas; they are stable at low, intermediate and high temperatures, respectively. For example, water, nitrogen and iron all behave in this way on heating, although at room temperature each has reached a different stage in the progression. Intermediate phases exist, including glasses, rubbers, liquid crystals, superconductors and superfluids. Each of these states of matter is characterized by a different degree of order in its internal structure; i.e. the regularity of arrangement of the individual atoms varies. The degree of mobility of the individual atoms and molecules also varies; the higher temperature states being characterized by a loss of order, and increased energy and mobility (Table 3.1).

Within this framework the crystal is the lowest energy, lowest entropy (highest regularity or lowest disorder) stage. The molecules are arranged in an ordered, symmetrical fashion so as to maximize intermolecular bonding which results in them being tightly packed in regular, repeating units throughout the crystal. This follows because in order to maximize bonding the molecules must be as close together as is possible and consistent with whatever directional requirements are imposed by the bonds. High-density packing of large number of similar objects can only be achieved by regular arrangements, otherwise large holes remain. Thus the structure and properties of crystals all follow from this principle of energy minimization.

On melting, the energy and mobility of the molecules increases, intermolecular bonding weakens, and the material enters the higher entropy liquid state, with a more randomized structure. Further heating leads to a loss of all order and the molecules fly around in a high energy, random fashion without any relationship to each other–the gaseous state.

23

Table 3.1   The different states of matter

State	Degree of order (regularity of arrangement)	Molecular mobility
Crystal	High	No
Liquid	Short range only	Yes
Gas	None	Very high
Glass	Short range only	No ('supercooled liquid') ('amorphous solid')
Rubber	Short range only	Yes, short range (Cross-linked macromolecular liquid)
Liquid crystal	High in one dimension Liquid-like in two dimensions	Yes, in two dimensions
Superfluid	Long range only	Yes (liquid helium)

3.1.2   Internal symmetry

When it was discovered that a crystal in a narrow X-ray beam gave a pattern of discrete spots, Bragg realized that the molecules within the crystal could be thought of as forming planes separated by a distance similar to the wavelength of X-rays (0.1–1 nm) (Bragg, 1961). The result is a three-dimensional equivalent of a diffraction grating for light (about 10 000 lines cm^{-1} or 1 μm separation). It is possible to determine the position of every atom in the crystal by analysis of the pattern of spots from a single crystal, usually recorded now by an electronic detector which can be positioned at any angle to the crystal and to the X-ray beam by a computer-controlled drive system. The first stage in this process is to determine the symmetry properties of the crystal which are also reflected in its external shape. This is done by assigning the symmetry of the X-ray pattern to a lattice and point group. Consider the analogous patterns (in two dimensions only) on a wallpaper of which the pattern is a group of four identical cats repeated at regular intervals. The lattice symmetry corresponds to the repeat of the four cats motif on a square, triangular or hexagonal grid whilst the point group describes internal symmetry within the four cats motif, for instance they might be in two pairs going opposite ways. Having established the overall (space group) symmetry of the arrangement one can determine all the dimensions of the molecule (or cat) from the intensities of the individual diffraction spots. Fig. 3.1 summarizes these elements of crystal symmetry. In order to do such a full analysis it is necessary to start with a perfect single crystal of about 1 mm^3. In powder diffraction, as would be used on samples

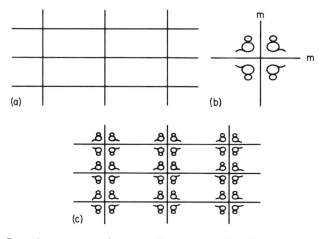

Fig. 3.1 Crystal symmetry in two dimensions: (a) a lattice, e.g. primitive rectangular; (b) plus a suitable point group, e.g. two mirror axes, m.m., and a motif (cat); give (c) a final structure-plane group p2 mm. In three dimensions fourteen possible Bravais lattices and 32 point groups give rise to 230 space groups.

from a diseased joint, a powder of small crystals gives a series of circles, usually recorded as arcs on a strip of film. Having only a dozen or so lines to work with, as opposed to thousands of spots, the structure cannot usually be determined directly but there is enough information to serve as a 'fingerprint' for the substance.

The 32 point groups of crystal symmetry described in Fig. 3.1 fall into seven fundamental classes. These seven classes are described in Fig. 3.2. They are closely related to both the internal structure and external form of crystals, and the terms are commonly used to describe the basic symmetry and type of a crystal.

One chemical may occur in several crystalline forms with different symmetries of molecular arrangement, and thus different space groups. Calcium pyrophosphate dihydrate is an example; it can form both monoclinic and triclinic crystals.

Not all solids are crystalline. Glass is an example of a solid lacking the symmetrical arrangement of molecules that characterize a crystal. Glasses are formed by rapid cooling, and the atoms do not have time to arrange themselves in an ordered fashion with the tightest possible packing and lowest energy. Solids with this more random arrangement are called 'amorphous', and will not produce the sharp, well-defined X-ray diffraction pattern that characterizes crystals since this depends on the presence of symmetry, and repetition of the same intermolecular spacing (Fig. 3.3). Crystalline substances also vary in their grain size. Many minerals contain

System	Crystallographic elements	Number of point groups	Appearance
Cubic	Three axes at right angles; all equal	5	
Tetragonal	Three axes at right angles; two equal	7	
Orthorhombic	Three axes at right angles; unequal	3	
Monoclinic	Three axes, one pair not at right angles; unequal	3	
Triclinic	Three axes not at right angles; unequal	2	
Hexagonal	Three axes coplanar at 120°; equal	7	
Trigonal	Three axes equally inclined, not at right angles; all equal	5	

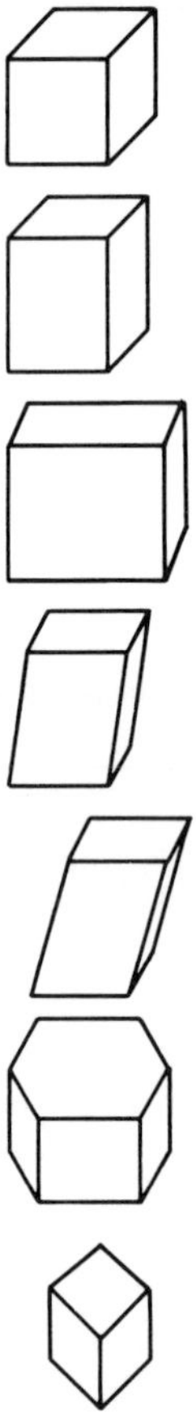

Fig. 3.2   The seven crystal classes.

individual crystals which can be seen by the naked eye and some mineral crystals are very large. The most valuable diamonds are the biggest ones, although small, microscopic diamonds share the same properties, such as their hardness, making them useful in industry. (Thus the distinction between 'industrial' and 'gem' diamonds). If the individual crystals of a mineral are very small or contain many defects the X-ray diffraction pattern becomes less sharp although it does not go to the very broad diffuse pattern characterizing the amorphous state.

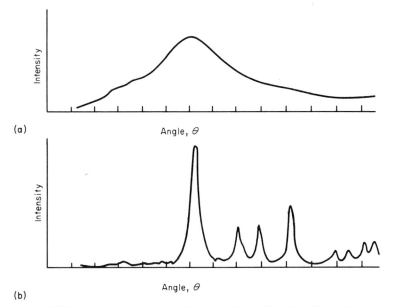

Fig. 3.3 Diffraction patterns from (a) amorphous (fused) silica, (b) crystalline silica (cristobalite). (Reproduced with permission, Uhlmann, 1971.)

3.1.3 Properties of crystals

The tight, regular bonding within crystals confers a variety of important properties (Table 3.2).

The closeness of packing results in increased density, and crystals are normally about 10–20% more dense than the corresponding liquid. Exceptions are water and bismuth, where the liquids are more dense than the crystals at the freezing point. In water the geometric constraints imposed by each molecule forming four hydrogen bonds make the ice crystal 10% less dense than water where hydrogen bonding is incomplete.

The hardness of crystalline materials reflects the difficulty of displacing the molecules relative to one another. Covalently bonded crystals such as silica and diamond are very hard since movement of the atoms actually involves breaking the highly directional bonds. On the other hand salts and metals are relatively soft since the bonds are not restricted to a single direction and the atoms or molecules can move without them being totally broken. However, crystalline materials are much harder than other biological materials and so can readily penetrate, abrade or indent them.

Dislocations are important in the understanding of the hardness and derformation of crystals. The forces necessary to distort a copper bar are far

Table 3.2   Some important properties
of crystals

Consequences of close packing

High density
Hardness
High refractive index

Consequences of regular packing

Cleavage on symmetry planes
Facetting on fracture
Facetted growth morphology
Sharp X-ray diffraction patterns
Optical birefringence
Piezoelectricity (sometimes)

---

lower than would be expected for the relative displacement of sheets of atoms in the crystal. This is because deformation occurs by movement of these line defects through the crystals so that the atoms move only one line at a time. As an analogy consider a ruck (hump) formed parallel to one edge of a carpet. By careful manipulation the ruck can be moved across the carpet in such a way that when it emerges at the opposite edge, the whole carpet has become slightly displaced. At any time only one small region of the carpet is in motion. Thus in terms of a distorting crystal, only a small number of bonds along a single line are disturbed at any time.

The cleavage and fracture properties of crystals also reflect the regular arrangement of the atoms. Because the bonds will be weaker along some molecular planes they tend to fracture more easily along some directions. Hence if a gemstone is struck nearly parallel to a weak plane it will fracture to reveal a facet, thus gems can be 'cut'. These easy fracture planes always bear a simple relationship to the symmetry of the unit cell. In the same way if a broken material has some small, absolutely flat surfaces it is crystalline.

The optical properties usually associated with crystals are the colour, sparkle and lustre of jewellery. Colour is essentially a molecular property since the ability of a material to absorb some or all frequencies of light is more a function of its chemical structure than of the relative arrangement of molecules. In this regard diamond and graphite are the exceptions which prove the rule. On the other hand refractive index is a property which depends both on chemical structure and density and so is a property that varies between different crystalline forms of the same substance and varies with direction within crystals. One consequence of this directional variation is that polycrystalline aggregates, such as chalk or sugar lumps, appear opaque and white. As the light passes from crystal to crystal it encounters changes in refractive index which scatter the light back out. The directional dependence of refractive index also causes crystals to rotate the plane of polarization of light. This birefringence is widely used to detect crystals in biological media as discussed in Chapter 4.

3.1.4   Crystals in joints

A variety of different crystals have been identified in joints (Tables 2.1 and 3.3). Most are derived from purine metabolism (such as monosodium urate monohydrate), or are calcium phosphates. However, a variety of lipids and other substances may crystallize in joints, and other particles can find their way into joint cavities. The crystals which grow to produce joint disease arise from body metabolites which are harmless in solution.

Joint crystals may crystallize *in vivo* (intrinsic crystals), or enter the body from outside (extrinsic crystals). In the next part of this chapter the principles behind the formation and growth of crystals, and the introduction of extrinsic crystals into the body, are considered. The formation of crystals in body fluids is a particularly complex situation; not only can a phenomenal number of structures be formed, but the solvent molecules may

Table 3.3   Varieties of joint particle

1. Form *in vivo*	Purine derivatives	Monosodium urate monohydrate
	Calcium salts	Calcium pyrophosphate dihydrate
		Hydroxyapatite
		Dicalcium phosphate dihydrate
	Others	Cholesterol
		Calcium oxalate
2. Derived from the joint	Bone and cartilage fragments	
3. 'Extrinsic'	Injected crystalline steroids	
	Foreign bodies	
	Plant thorns	
	Fragments of prostheses or their cement	

also be taken into the lattice during crystallization. Thus a solution may crystallize an unhydrated crystal (e.g. salt, NaCl) or hydrated structures (e.g. copper sulphate, $CuSO_4 \cdot 5H_2O$). These complex hydrated crystals are also hard, and can have an active surface capable of inducing inflammation and other pathological reactions.

3.2   The origin of 'intrinsic' crystals

3.2.1   Phase changes

Most pure, simple compounds progressively change from solid to liquid to gas when heated. In simple terms one can think of this as the thermal vibrations progressively overcoming the attractive forces between molecules as the temperature is raised. Thus they go from the regularly arranged rigid crystalline structure to the less regular but still closely packed mobile liquid state and then to the total independence of the gaseous state. What is not obvious according to this picture is why we have only three states with abrupt transitions between them rather than a smooth, continuous transition from crystal to gas. Indeed at sufficiently high pressures the dividing line between liquid and gas is lost but melting and freezing are always sharp transitions which give rise to new surfaces during the phase change.

In order to be able to discuss phase changes we need to be able to talk in thermodynamic terms. The key function is the free energy, G, measured in the same units as heat, calories per gram or joules per kilogram. This is a composite of the enthalpy, H, and entropy, S, which both depend on the structural state of the compound. The enthalpy is a measure of the bonding strength between the molecules in that strong bonding corresponds to a low value of enthalpy. Entropy measures the disorder in the substance. Thus in a gas where the molecules are flying about at random the entropy and enthalpy are high whilst in a crystal where the molecules are rigidly fixed into regular positions both the entropy and enthalpy are low. Free energy, then, is defined by the equation:

$$G = H - TS \qquad (3.1)$$

where T is the temperature in degrees Kelvin ($0°C = 273$ K). The normal, stable state of a substance at any temperature is that which has the lowest free energy. Thus from Equation (3.1) we can see that a material will tend to minimize its enthalpy at low temperatures but maximize its entropy at high temperatures. This is shown schematically in Fig. 3.4 where the free energy of liquid is initially above that of crystal but crosses it as the temperature increases because the slope is steeper, i.e. the entropy is larger.

Before moving on from pure compounds to solutions and mixtures, two

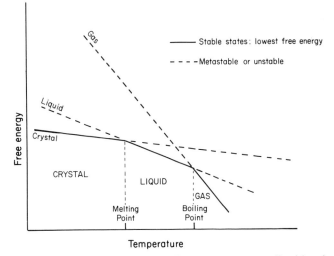

Fig. 3.4   Schematic change of free energy with temperature: gas, liquid and crystal.

other points should be repeated. There are frequently more than three states of a substance. For instance elongated molecules tend to form liquid crystals which appear in a narrow temperature range around the crystal-to-liquid transition and would be characterized by a line of intermediate slope as shown in Fig. 3.5. Although these complicate our simple picture they are nonetheless well-defined states separated by sharp transitions. Similarly there may be several crystal structures of a compound which are stable at

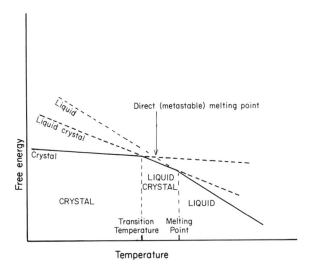

Fig. 3.5   Schematic free-energy diagram: liquid, liquid crystal and crystal.

different temperatures. Hence we have graphite and diamond as two forms of carbon. Graphite is the stable form at room temperature although, perhaps fortunately, diamond does not convert to graphite at a detectable rate. Similarly there is a transition from metallic beta to brittle alpha tin which occurs at low temperatures and caused Napoleon's army to lose its buttons in the Russian winter.

It is most important for our subsequent discussion to realize that as for diamond and graphite, thermodynamic diagrams can define the equilibrium transition points between phases but not whether those transitions will actually occur in a finite time. Thus a liquid may be heated by several degrees above its boiling point, following the liquid free energy line in Fig. 3.4 rather than changing to the gas line. To avoid this and the sudden explosive boiling that follows, chemists add anti-bumping granules to stimulate the transition. Water in droplet form may be cooled to $-40°$ C without freezing, well below the water-to-ice transition temperature. Cloud-seeding experiments seek to add nucleating agents to speed up the transition and consequently initiate precipitation. If a liquid is cooled sufficiently far without freezing, usually to about two-thirds of its absolute (Kelvin) melting temperature, it goes through a glass transition to an amorphous solid state without any sharp change in enthalpy or entropy as shown in Fig. 3.6. This is a common occurrence in many viscous liquids such as glycerol, oils and molten inorganic oxides like silica and glass. The glassy state has no more order than a liquid, which means there is no regularity detectable by X-ray diffraction over a range of more than a few molecules. It has a higher free energy than the crystal and therefore is a metastable state, it would convert to crystal if a mechanism existed for doing

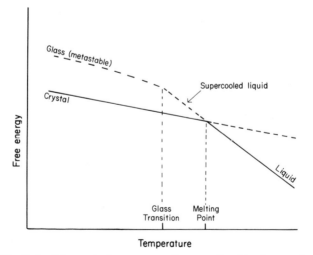

Fig. 3.6 Schematic free-energy diagram: liquid, glass and crystal.

so. In this context it will be understood why old glass is sometimes found to have spherical crystalline inclusions.

Thus for a phase change to occur there must be a decrease in free energy taking the system closer to equilibrium but there must also be an easy route available for the material to undergo the transition in a reasonable time. This relationship between stable, metastable and unstable states is illustrated in Fig. 3.7.

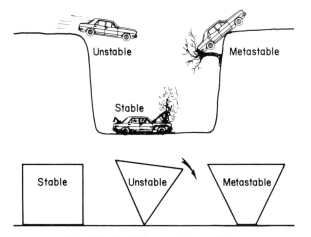

Fig. 3.7 Stability and instability.

3.2.2 *Solutions*

The growth of crystals from solutions is in many ways analogous to the freezing of pure liquids; although solution concentration is an additional controlling variable as well as temperature, so that saturation solubility plays the role of melting point. However, the fact that we now have at least two chemical species present, solute and solvent, means that there are many more possibilities for complications. The important state function is now the chemical potential of the solute, μ_b (conventionally the solvent is 'a' and the solute 'b'). This is the free-energy change of the solution when a small amount of solute is dissolved in a large volume of solution such that the concentration is not significantly changed. It bears the same relationship to free energy as does marginal cost, the cost of producing one extra of an item, bear to the average cost of a product in economics.

If a solution of a crystalline solid behaves ideally, i.e. the solute and solvent molecules are indifferent to one another in solution and there are no preferential interactions, the chemical potential varies with concentration as shown in Fig. 3.8. There is a balance between two terms. The first term is

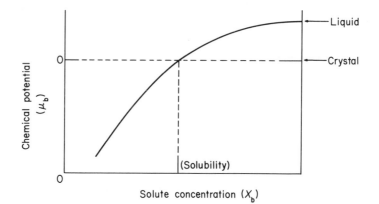

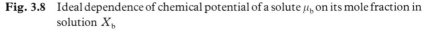

Fig. 3.8   Ideal dependence of chemical potential of a solute μ_b on its mole fraction in solution X_b

$\mu_b = RT \ln(X_b) + \Delta H_f (1 - T/T_m)$

R = gas constant, T = temperature, T_m = solute melting point, ΔH_f = solute heat of fusion. In water $X_b \simeq$ solute molarity/55.

the free energy decrease (entropy increase) due to the mixing between solute and solvent. This is the increased randomness caused, for example, by mixing a few pins into a box of needles. The effect is large for the first few pins added but the rate of change becomes less with more additions. This is also why purification of gases or solvents gets progressively more difficult as higher levels of purity are reached. The second term is the free energy increase of the solute on going from the tightly bonded crystalline state to a looser, liquid state. Crystals will dissolve until the mixing term is no longer large enough to outweigh the second term. The chemical potential of pure solute crystals is taken as zero so the equilibrium solubility is defined by the point at which the chemical potential of the solute in solution, μ_b, also becomes zero.

We can use this to calculate the solubility as a function of temperature as shown in Fig. 3.9.

In fact this simple scheme is rarely followed, especially in aqueous solutions where the interactions tend to be very strong. However, it is often possible to treat solubility in terms of deviations from ideality so that these relationships are a good guide to the type of behaviour to be expected.

In addition to temperature a wide range of other factors influence the solubility of a compound. If a substance has several different crystal forms the metastable ones will be more soluble than the stable form. Similarly glassy materials are more soluble than the equivalent crystals. Ideal solution theory predicts that they would be completely soluble as they are already in an amorphous state so that our second, energy, term is zero.

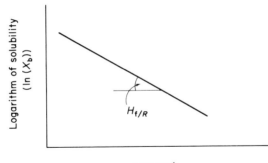

Fig. 3.9 Expected variation of solubility with temperature from the ideal solution equation of Fig. 3.8.

$$\ln X_{\mathrm{b}} = (\Delta H_{\mathrm{f}}/R) \left(\frac{1}{T_{\mathrm{m}}} - \frac{1}{T} \right)$$

Glassy ('fused') silica has a solubility in water of 100 ppm while quartz, chemically identical but crystalline, has a solubility of 6 ppm. Solubility also increases with decreasing crystal size due to the effect of the increased surface energy but the effect is very small except for submicroscopic crystals. However, the increased surface area does mean that very small crystals dissolve much faster than larger ones.

3.2.3   Aqueous solutions

In aqueous solutions of ionic compounds the binding of water molecules to the dissociated ions gives rise to effects so strong that there is little point in discussing them in terms of deviations from the ideal; we have to start from a new reference point. The details of this treatment do not belong here, the reader is referred to a text on chemical thermodynamics (Moore, 1972).

The result is the solubility product, a constant defining the saturation condition:

$$P = a_{\mathrm{A}}{}^{x} a_{\mathrm{B}}{}^{y} \quad \text{at saturation} \tag{3.2}$$

where a_{A} and a_{B} are the activities in solution of the A and B ions of the compound $A_x B_y$. In very dilute solutions the activities of the ions are just equal to their molar concentrations and the solubility product is then a product of ionic concentrations. Table 3.4 gives some values for the activity coefficient γ, which is the ratio of activity to concentration, as a function of solution concentration. It can be seen that γ is almost 1 in dilute solutions, but as the ionic strength rises pairs of ions of unlike charge tend to attract one another in solution, so reducing their effective activity so that the activity

Table 3.4 Activity coefficients, γ, of a number of solutes at different concentrations in water

	Concentration (M)			
Solute	10^{-3}	10^{-2}	10^{-1}	1
NaCl	0.966	0.904	0.780	0.66
$CuSO_4$	0.74	0.41	0.16	0.047
$CaCl_2$	0.89	0.725	0.515	0.71
Sucrose	1	1	1	1.15

An activity coefficient of 1 means that concentration can be used in place of activity when determining solubility. In physiological saline the correction factor is about 0.75.

coefficient is less than 1. The effect is much stronger with multiply charged ions like copper and sulphate. Non-ionic compounds like sucrose lead to a slight increase in activity coefficient as they bind up some of the water and reduce the effective amount of solvent. The deviation between activity and concentration is significant in physiological saline (0.14 M). Theories of these effects are not good except in very dilute solution, but there is a great deal of empirical data which can be used in calculations on biological systems.

Supersaturation occurs when for one or more of the pairs of ions in solution the product on the right-hand side of Equation (3.2) exceeds their respective solubility product P. The supersaturation ratio can then be defined as the ratio $(a_A{}^x a_B{}^y / P)^{1/(x+y)}$, the $1/(x+y)$ power being necessary so that ionic compounds are counted in the same way as non-ionic compounds. Alternatively we can subtract one from this ratio to obtain a supersaturation which increases from zero at equilibrium. Since we will not always want to distinguish between ionic and non-ionic compounds we will identify the product $(a_A{}^x a_B{}^y)^{1/(x+y)}$ with C the effective concentration of solute and $P^{1/x+y}$ as S the saturation concentration. Thus the supersaturation is $(C-S)/S$ and if the activity coefficient does not change much in the concentration range of interest we can just work with the product of the ionic concentrations.

From Equation (3.2) we can also derive the common-ion effect where, for instance, a subsaturated solution of copper sulphate can be induced to precipitate by addition of sodium sulphate so that the product (copper) times (sulphate) is increased. Fig. 3.10 shows the effect of adding salt to a solution where (a) the solute is ionic but there is no common ion so that we just see the effect of ionic strength on activity coefficient, γ (b) the solute has a common ion with the added electrolyte so that this effect is superimposed on

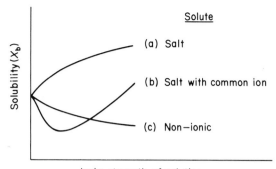

Solute

Ionic strength of solution

Fig. 3.10 Effect of changing the ionic strength on the solubility of (a) a salt which has no common ion with that in solution, e.g. $CuSO_4$ in NaCl solution, (b) a salt which has a common ion with that in solution, e.g. $CuSO_4$ in Na_2SO_4 solution (the common ion initially depresses the solubility). (c) a non-electrolyte, e.g. cholesterol in NaCl solution.

that in (a), and (c) the solute is non-ionic and the added electrolyte binds some of the water leaving less free to mix with the solute.

For sparingly soluble solutes such as uric acid we may have to consider several possible precipitates before deciding whether a solution is saturated. Thus from a solution of uric acid and sodium chloride in water we may obtain uric acid, monosodium urate or disodium urate or two of these depending on the concentrations and pH. Fig. 3.11 shows the conditions

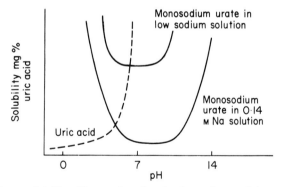

Fig. 3.11 Urate solubility. Curves showing the dependence of the solubility of uric acid and sodium urate on pH and sodium concentration. At low pH all the urate in solutions exists as uric acid. Above pH 5 progressively more of the urate exists as the singly charged anion, so uric acid is more soluble. However, the presence of urate ion increases the tendency of monosodium urate to precipitate, dependent on the sodium concentration. Above pH 9 there is no uric acid but increasing amounts of the doubly charged urate anion and monosodium urate again becomes more soluble (after Wilcox *et al.*, 1972).

under which uric acid and sodium urate will precipitate from solution in water or in 0.14 M sodium chloride as a function of pH. The solubility of uric acid is much less than of urate because it is not stabilized by the strong interaction between water and charged species.

3.2.4   Driving force for crystallization

It is well known that supersaturated solutions frequently do not crystallize although crystals in subsaturated solutions always dissolve. Thus Fig. 3.12 shows the temperature dependence of solubility and of supersolubility, the concentration above which precipitation will start spontaneously. Although, as implied by this, crystallization often appears to start quite abruptly at a particular temperature when a saturated solution is cooled; the position of this supersolubility line is dependent on the cooling rate, the volume of the solution and on the impurities present. Thus it cannot be regarded as a constant property of the solute but as a useful practical guide to precipitation.

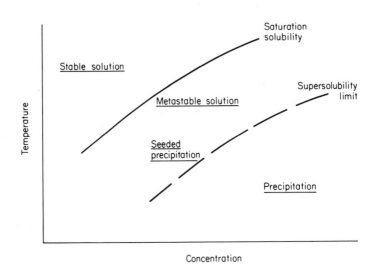

Fig. 3.12   Regions of stability for a solution.

Once crystallization has started it will continue until the solution concentration is reduced to saturation. Hence 'seeding' with small crystals can be used to initiate precipitation in the metastable region.

If we look at Fig. 3.8 we can see why there might be an increase in the tendency to crystallize as we move away from saturation, in that the chemical potential difference between the dissolved and solid states increases and so the 'driving force' for crystallization increases. This driving force is the difference in chemical potential of the solute between the supersaturated solution and the crystals of solute. We can get this by writing the ideal solution expression for chemical potential and subtracting the saturation value $(\mu_{b,s})$ from the supersaturated value $(\mu_{b,c})$. The result is shown in Fig. 3.13. For small supersaturations this gives a driving force which is proportional to supersaturation $(C-S)$:

$$F = RT\,(C-S)/S \tag{3.3}$$

Thus the driving force for crystallization depends on the logarithm of the supersaturation ratio or, at small supersaturations, on the supersaturation itself. For an ionic compound these ratios would involve ion products rather than concentrations.

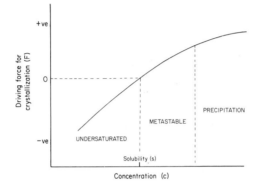

Fig. 3.13 Thermodynamics of crystallization.
Driving force $F = (\mu_{b,c}) - (\mu_{b,s}) = RT\,\ln\,(c/s)$.

3.2.5 Solubility in biological fluids

Biological fluids such as blood or synovial fluid are not different in principle from simple solutions but they do tend to be complex, variable and unstable mixtures. The conditions of pH $(7{\cdot}4)$ temperature $(30–37^\circ\ \mathrm{C})$ and ionic strength $(0.14\ \mathrm{M}\ \mathrm{NaCl})$ are quite well defined though there may be important variations due to local inflammation, disease or trauma. There are problems in defining the concentrations of minor components such as uric acid or phosphates and their activity may be affected by binding to blood or tissue proteins or proteoglycans. In addition liquids like blood contain cells and lipid phases which are not necessarily in equilibrium with the aqueous phase and may control the behaviour of water-insoluble compounds like cholesterol. These factors make it difficult to establish the true activities of crystallizing metabolites. They can really only be resolved by a careful analysis of each separate case.

Gels, such as articular cartilage, are even more complicated, although often the initial site of crystal deposition. As outlined in Chapter 2, the complex organization of macromolecules like proteoglycans can lead to preferential diffusion, and binding of certain ions, and thus 'partition effects' altering the effective concentration of a solute in cartilage. Calcium concentrations have been studied, and vary with different depth of cartilage, but the activity of many other solutes remains undetermined. In general most of these types of interactions can be determined or estimated so that biological systems are complicated but not intractable.

Having established for any system that it is supersaturated it remains to decide whether or not crystals will grow. In the next sections we will discuss the factors affecting crystal growth rates.

3.2.6   Growth kinetics

Fig. 3.14 illustrates the progress of crystallization of a supersaturated solution of monosodium urate at 37° C. The curve can be divided into an induction period, a growth phase and a slow approach to equilibrium. The induction time is dominated by the nucleation and growth of small crystals which start to measurably deplete the solution in the growth phase. Finally the solute concentration becomes so close to equilibrium that the growth rate decreases towards zero. Seen in terms of individual crystals there are two basic processes, the formation of crystal nuclei and the growth of crystals once formed.

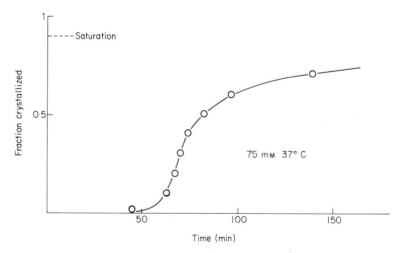

Fig. 3.14   Typical development of crystalline precipitate with time, for a solution of monosodium urate.

In Fig. 3.14 we have information on the nucleation rate, initial growth rate and the decrease in growth rate as the concentration approaches saturation. It is tempting to identify the initial flat part of the curve, the induction time, with a time for nucleation but a similar lag time would also be seen if a number of nuclei were formed instantly because until the crystals are quite large their increase in radius removes little volume of solute. We can readily fit curves such as Fig. 3.14 given suitable expressions for the nucleation and growth rates but we should be wary of identifying one part of the curve with one process alone.

3.2.7   Nucleation

Fig. 3.15 shows the driving force for crystal growth as a function of the size of the crystal. It can be seen that when it is very small (a few tens of molecules), this driving force is negative, meaning that molecules will be more likely to leave the crystal than add to it. This arises from the fact that a very small solid has a large ratio of surface area to volume. Thus the crystal-to-liquid surface energy is initially more important than the internal binding energy of the crystal. However as the crystal grows the surface area increases as the square of the radius whilst the volume increases as the cube of the radius and so the surface energy ceases to be important relative to the volume energy. From Fig. 3.15 we see that there is a critical size which a crystal must exceed in order to grow. This size can only be reached by a chance fluctuation against the prevailing free-energy gradient and so if the critical nucleus is large as it is at small supersaturations, crystal nucleation is most unlikely. As the supersaturation increases the critical nucleus becomes smaller until they are formed at measurable rates. Hence as shown in Fig. 3.16 nucleation in the absence of foreign particles, is extremely slow at low supersaturation but increases rapidly at higher concentrations. In fact the supersaturation dependence of nucleation is so strong that it goes from

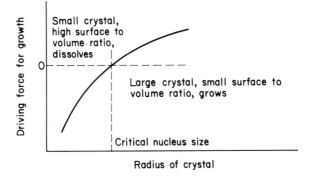

Fig. 3.15   Driving force and crystal size.

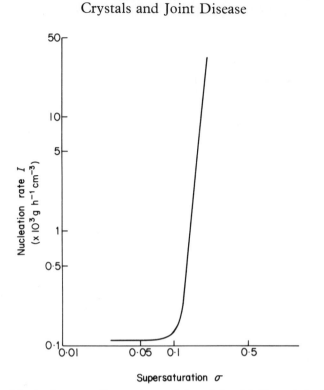

Fig. 3.16   Nucleation rate of potassium sulphate (after Mullin and Gaska, 1969).

being immeasurably slow to immeasurably fast over a very narrow range corresponding to the supersolubility line. In practice most solutions contain particles of dust which may be able to enhance nucleation by forming a low-energy boundary with the nucleus so that the critical radius is reached when the nucleus contains fewer molecules as shown in Fig. 3.17. In this sense the ideal seed is a crystal of the compound itself as the boundary energy will then be zero. If the seed particle is crystalline as it usually must be, it is thought that there has to be a specific relationship between the crystal lattice of the seed and that of the solute. The solute atoms then attach to the surface of the seed in a regular array close to that which they will form in the crystal. The phenomenon whereby one compound crystallizes on the surface of another with a specific orientational relationship between the two crystals is called epitaxy.

Homogeneous (dust-free) nucleation theory is well supported by experiments on fine liquid droplets forming from vapours and on crystallization of fine droplets of melts or solutions. In these experiments the fine dispersion of the liquid ensures that there are many more droplets than dust particles so that most drops will be free of dust and must nucleate without its aid.

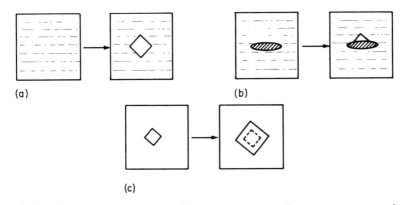

(a)   (b)

(c)

Fig. 3.17 Nucleation processes. (a) Homogeneous: a nucleus appears as a result of the random collisions of enough molecules to form a seed larger than the critical size. (b) Heteregeneous: the critical nucleus size is smaller when the seed is attached to a surface with which it has a low surface energy. If the growing crystal has a fixed orientation to the substrate this is called epitaxy. (c) Seeding: crystals of greater than the critical size are introduced; nucleation is thus circumvented.

Relatively little is known about heterogeneous (contaminated) nucleation from aqueous solutions as the subject is hard to study experimentally, but effective nucleating agents seem rare. Nucleation can often be induced by scratching the vessel wall or by ultrasonic irradiation. These may have the effect of detaching small pre-existing crystals from the vessel wall and breaking them up into many fragments. Alternatively fresh glass surface may be an active nucleating agent. Nucleation frequently occurs at the air–solution interface possibly due to drying in solution films left on the wall.

A great deal of effort has gone into investigating the role of collagen as an epitaxial nucleating agent for hydroxyapatite or its precursors (see also Chapter 8). There is evidence for some forms of collagen being active though often at conditions close to those where spontaneous precipitation would occur in any case. It is not clear whether this is important for *in vivo* mineralization.

Similarly, the role of other biological 'surfaces' in crystal nucleation is unclear. Cell membranes, gel–sol interfaces, and macromolecules other than collagen could all theoretically provide suitable nuclei for crystal formation. Furthermore, a crystal of hydroxyapatite could nucleate growth of other compounds, such as calcium pyrophosphates. As discussed in Chapter 5, examination of crystals by electron microscopy lends some support to the idea that urates and other crystals may sometimes grow around a contaminant 'nucleus'.

Because nucleation occurs at high supersaturations it is common for several different crystal phases to be supersaturated at once. For instance in calcium oxalate solutions, $CaC_2O_4 \cdot H_2O$, $CaC_2O_4 \cdot 2H_2O$, $CaC_2O_4 \cdot 3H_2O$ may all be supersaturated to different extents in a metastable solution. Which phase then forms may depend on for which there are active nucleating agents and on other solution conditions. If a metastable phase forms first it may subsequently convert to a more stable form. As discussed in Chapters 5 and 6, calcium phosphates have many crystalline forms, and stable crystals such as triclinic calcium pyrophosphate dihydrate and hydroxyapatite can emerge via many metastable intermediates.

3.2.8 Growth

Once the crystal has become much larger than the critical radius in Fig. 3.15 the free-energy change per added molecule becomes constant and, in the absence of changes in the solution concentration, the crystal will grow at a constant rate. In principle one might expect molecules to attach to the crystal surface wherever they happen to collide with it with a probability of attachment per collision proportional to the driving force for crystallization. This would result in spherical crystals whose radius increased uniformly with time at a rate proportional to $\ln(C/S)$. Of course crystals are not usually spherical as addition occurs at different rates on different crystallographic faces such that characteristic faceted crystal shapes are maintained. Fig. 3.18 shows how the growth rates of some real crystals depend on supersaturation. The cause of the non-linear dependence of growth rate on driving force

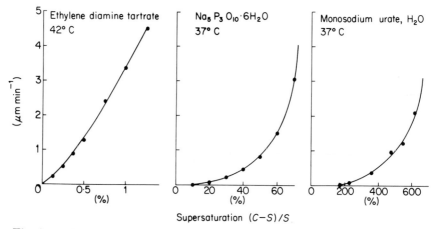

Fig. 3.18 Experimental variation of crystal growth rates from aqueous solution. Note huge variation in supersaturation ranges. (Booth and Buckley, 1955; Troost, 1972.)

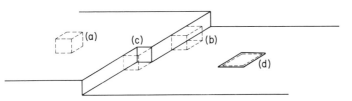

Fig. 3.19 Types of site on a crystal surface to which a molecule may add: (a) isolated atom, low binding energy; (b) adjacent to a step; (c) at a step site, strong binding, repeating; (d) in a hole, strong binding, self-terminating. Types (b) and (c) will be responsible for most of the additions taking place. Hence steps are necessary for rapid growth.

shown in Fig. 3.18 is that single molecules cannot attach to a flat crystal surface, growth can only occur by addition at step sites as shown in Fig. 3.19 where an added molecule is bound into the crystal on at least half its sides. Furthermore, the site must be such that addition of a molecule results in the production of a further similarly favourable addition site so that the process is self-propagating. The equilibrium state for the surfaces of most crystals in contact with solution is perfectly flat so special mechanisms are necessary to produce such growth steps. They are believed to arise at crystal defects, usually screw dislocations emerging at the surface, or as a result of the spontaneous addition of a cluster of molecules in a surface, two-dimensional analogue of the nucleation process. These mechanisms are illustrated in Fig. 3.20. The dislocation-controlled growth process leads one to expect a

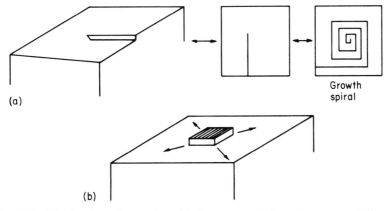

Growth
spiral

(a)

(b)

Fig. 3.20 Production of step sites. (a) At a screw dislocation: screw dislocation emerging from a surface. Addition of molecules to the step during growth causes it to wind into a spiral. (b) By nucleation on flat surface: a small disc of atoms attach spontaneously to the flat surface. The disc grows if it is larger than a critical size.

growth rate proportional to the square of the driving force whilst the surface-nucleation model predicts an exponential dependence. These relationships are illustrated in Fig. 3.21 and it can be seen that they do resemble actual growth data.

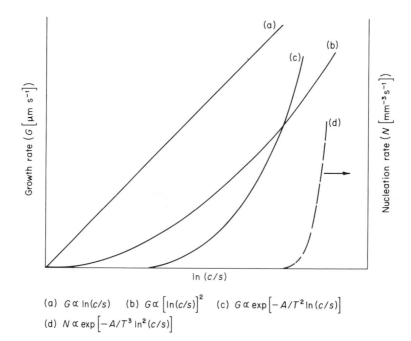

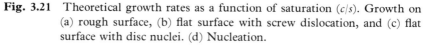

(a) $G \propto \ln(c/s)$ (b) $G \propto \left[\ln(c/s)\right]^2$ (c) $G \propto \exp\left[-A/T^2 \ln(c/s)\right]$

(d) $N \propto \exp\left[-A/T^3 \ln^2(c/s)\right]$

Fig. 3.21 Theoretical growth rates as a function of saturation (c/s). Growth on (a) rough surface, (b) flat surface with screw dislocation, and (c) flat surface with disc nuclei. (d) Nucleation.

These models are very simplified and further complexities are added by the fact that molecules may first adsorb to the crystal surface and then diffuse toward the growth step as well as by many more details of the addition process. There is a substantial amount of evidence for the dislocation mechanism. Several systems have been shown to follow the square-law driving force dependence, including calcium sulphate, calcium oxalate and potassium sulphate (Nancollas, 1979) and growth spirals have been observed on crystal surfaces although the fact that they are observable means that the steps are about one hundred molecules high rather than just one crystal plane: this is probably due to the action of impurities in causing moving steps to bunch like cars on a motorway. We have found evidence for exponential surface-nucleation growth kinetics in aqueous solutions of sodium urate and two proteins, lysozyme and insulin. In all these cases quite high supersaturations were used. Probably all systems will follow square-law growth kinetics at very low supersaturations where nucleation is infinitesimally slow. Consequently growth in this regime will be very sensitive to the presence of dislocations in the crystal and there will be large variations from crystal to crystal. At higher supersaturations nucleation-controlled growth is likely to take over as it becomes reasonably fast. Eventually supersaturations may be so high that the surface is covered in growth steps and the rate becomes proportional to driving force. Recent computer simulations of crystal growth by Gilmer (1980) have shown these relationships between the various growth mechanisms in a way that is more convincing than the rather arbitrary arguments given above.

Since these processes are dependent on surface energies it is not surprising that they depend on the molecular configuration at the surface and hence that crystal growth rates vary greatly from one crystal face to another. In general the highest step site energies will be on those surfaces where the distances between successive crystal planes parallel to the surface, the step heights, will be greatest. These will be the simplest crystal planes and so the slowest growing faces with the highest step energies will reflect the simple crystal symmetry elements. As shown in Fig. 3.22 fast growing faces will tend to disappear from a growing crystal so crystals are usually

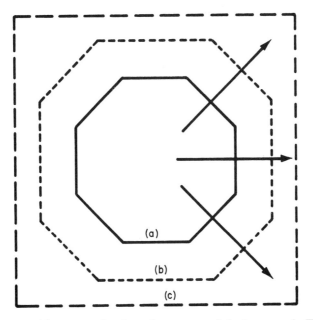

Fig. 3.22   Loss of faster growing faces from a crystal during growth. The crystal (a) has fast-growing corner facets which become smaller as the crystal grows (b), only the slower faces remain (c).

bounded by their simple slow-growing faces which directly reflect their
internal symmetry.

3.2.9 Poisons

Crystal growth poisons are substances which affect the growth and
dissolution rates of crystals when present in amounts much less than
that of the crystallizing solute. In other words they act on the crystal surface,
not by complexing the solute. A large number of examples are known but
their mode of action is usually unclear.

Buckley (1951) has described a wide variety of organic dyes which affect
the crystallization of inorganic salts at concentrations 1/1000 of the salt concentration. Growth rates are greatly reduced, dissolution rates are slightly
reduced and the habit, the crystal shape, is modified. The dye is taken into
the growing crystal and can often be seen to have adsorbed on specific
growth faces. Similarly Nancollas (1979) has shown that pyrophosphate at
concentrations down to 10^{-7} M reduces the nucleation and growth rate of
calcium oxalate monohydrate and of dicalcium phosphate dihydrate (Fig.
3.23). There is also a much smaller decrease in the dissolution rate. It is
thought that urine pyrophosphate might be an important factor in
preventing kidney stone formation (Chapter 10).

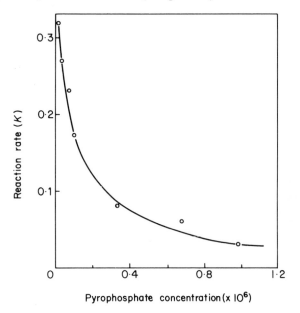

Fig. 3.23   The reaction constant (to which the rate of growth is proportional) for the
growth of calcium phosphate as a function of pyrophosphate concen-
tration (after Nancollas, 1979).

As a function of the poison concentration the growth-retarding effect follows a Langmuir adsorption isotherm. Since at 10^{-7} M there is too little poison to completely cover the crystal surface, it seems that the poison must specifically bind to step sites and prevent the further addition of solute. The effect can be quite selective, some poisons affect nucleation but not growth of calcium oxalate trihydrate and have no effect on calcium oxalate dihydrate. An interesting example of a biological growth poison is the glycoprotein present in the blood of antarctic fish to prevent ice formation at low temperatures. Another poisoning effect probably controls the formation of the aragonite form of calcium carbonate rather than calcite in some mollusc shells.

Following the finding that pyrophosphate could inhibit crystallization of hydroxyapatite, a number of stable organic phosphonates have been investigated. They affect a large number of anions, and can poison many crystals, including all organic calcium phosphates. Their use in crystal deposition diseases is therefore being explored, but is complicated by the fact that they have other independent effects on cell function (see Chapter 11).

Several polymers also act as crystal poisons, presumably by a surface-coating effect that inhibits growth. As explained in Chapter 2 the connective tissue of the joints consists largely of proteoglycan polymers, and there is some evidence that aggregated proteoglycan can inhibit the growth of urate crystals (Katz and Schubert, 1970), although more recent work suggests that this may be an artefact due to ion exchange of sodium with potassium bound to the proteoglycan preparation (Perricone and Brandt, 1978; 1979). Hyaluronic acid in synovial fluid also is a large polymer, and may help inhibit crystal deposition in joints. Natural polymers may well play an important part in inhibiting bone growth and preventing more crystallization, and connective tissue glycoprotein may have a similar role in natural inhibition of crystal formation.

Other polymers are now being investigated in water treatments to prevent scaling (deposition of calcium carbonate crystals). These compounds inhibit crystal growth and raise the calcium concentration required before significant scaling occurs (sometimes known as a 'threshold effect'). These agents also inhibit growth of a wide spectrum of other crystals and are being investigated for possible medical applications.

3.2.10 Diffusion effects

So far we have only concentrated on the rate of addition at the crystal–liquid interface which is a function of the driving force, the availability of step sites and the rate at which molecules at the interface orient themselves correctly to attach to the crystal. They may do this either at the step site or at some remote place on the surface from which they diffuse as an adsorbed molecule until they encounter a step site.

Growing crystals may also deplete the surrounding solution of solute so that the growth rate becomes determined by the rate at which ions or molecules can diffuse through the solution to the crystal. Clearly this is most likely to be important for a rapidly growing crystal in an unstirred, viscous solution.

As mentioned in Chapter 2, cartilage has special diffusion properties, dependent on applied mechanical forces, and varying for different ions. This may well be an important determinant of the site and type of crystal deposit forming in this tissue.

3.2.11   Dissolution

Dissolution is essentially the reverse process to crystal growth and in thermodynamic terms it is analogous with subsaturation replacing supersaturation as the driving force. In general one is interested in the dissolution of a pre-existing crystal in a solution so that nucleation is not necessary. Also step sites can be readily generated by loss of molecules from the crystal edges and corners so that these need not be formed by surface nucleation except in large crystals. Dissolution rate seems to normally vary as the first or second power of the driving force in those few systems that have been measured. Often the process is limited by the rate at which the solute diffuses away from the crystal surface and so it is dependent on the stirring rate. The loss of corners means that crystals in a subsaturated solution often rapidly become rounded and quickly regain their facets if the solution becomes slightly supersaturated.

The rate of dissolution of crystals *in vivo* has been measured in the case of calcium pyrophosphate dihydrate, by use of a radioactive label incorporated into the crystal lattice (Chapter 5).

3.2.12   Morphology

The shape, size and degree of aggregation of crystals is a function of the supersaturation at growth and the composition of the mother solution. In general, growth at low supersaturations will produce a few large, perfect crystals. Higher supersaturations or lower temperatures tend to give abundant nucleation leading to many small crystals. At high supersaturations crystals may also nucleate in new orientations on the faces of existing crystals, leading to aggregates and, in elongated crystals, to spherulites. Dyes and other poisons can also lead to habit modification (shape changes). Rinaudo and Boistelle have recently reported a detailed study of growth morphology in monosodium urate and uric acid.

3.2.13   *In vivo and in vitro studies*

The growth of crystals in joints or other sites *in vivo* will be dependent on the same factors of solubility, nucleation and growth rates as *in vitro*. The solubility is a quantity which can be determined for the *in vivo* conditions with reasonable certainty and so the degree of supersaturation may be known. However local variations in metabolite concentration do occur in the body and these may be important as may time fluctuations. Also it should be remembered that the body surface temperature varies over a range of 10° C or so from the trunk to the extremities.

Laboratory experiments usually extend over a day or two whilst diseases may take years to develop. Thus crystal deposition disease may develop at supersaturations less than those at which any precipitation is detectable in the laboratory.

Epitaxial effects may be important in enhancing nucleation *in vivo* but as yet there is little evidence that this is so. Poisons are certainly effective in reducing *in vivo* growth rates but the effect seems often to be common to a wide range of related substances, such as large anions poisoning oxalate growth so specific effects which vary much between individuals seem unlikely. Sporadic local concentration fluctuations may be much more important in initiating nucleation.

There is much that is poorly understood in a simple precipitation experiment and the body is a far from simple system. However, except perhaps where specific metabolic processes are involved as in bone growth (Chapter 8), the behaviour of the living system should be analogous to the *in vitro* one and governed by the same general principles.

3.3   Extrinsic crystals

As well as growing *in situ*, crystals and other hard particles may enter the body from the outside and then cause tissue damage. Table 3.5 summarizes the main possible routes of entry and effects.

The most important and most studied diseases are those of the lung. In particular the occupational diseases caused by inhalation of crystalline or other dusts, are very common and disabling. The repiratory tract has a range of mechanisms to prevent the entry of particles suspended in the air or to remove them. These include the nasal cilia which filter the incoming air and the mucociliary escalator. This upward movement of secretions on the walls of the respiratory tract carries away those particles which collide with the walls. Phagocytic scavenger cells also line the airways. However, dust particles of an effective diameter below about 3 μm may not be removed as air passes down the airways, and a proportion of them precipitate in the small airways and alveoli. Here they may be taken up by the macrophages, and can cause damage to the lung, especially pulmonary fibrosis. As in other crystal deposition diseases the resulting problem depends on the type of particles, the dose delivered, and the susceptibility of the host tissues. Long thin particles (fibres) seem to be particularly harmful, and can cause carcinogenic changes (pleural mesothelioma) as well as fibrosis. A fibre is less likely to precipitate from the inflowing airstream than an equiaxial particle of the same volume whilst the length makes it prone to cause cellular disruption after phagocytosis.

Table 3.5 Possible routes of entry of extrinsic particles

Routes of entry	Inhalation
	Ingestion
	Skin penetration
	Injections
	Operations
Activity	Acute inflammation
	Chronic inflammation and granuloma
	Mechanical damage
	Carcinogenicity

Ingestion of particles rarely causes a problem unless the individual piece is large enough to cause mechanical damage (e.g. blockage of the tract or perforation by swallowed sharp objects). Drinking water and many beverages do contain particles, including asbestos, and particulate matter has been implicated in some diseases including carcinoma of the stomach and inflammatory bowel disease. There is, however, no convincing evidence that these particles are important, and the alimentary tract seems particularly resistant to these types of disease.

Penetration of particles through the skin can cause localized inflammatory reactions. For example, slate miners have characteristic chronically indurated skin lesions due to small pieces of mineral lodging in the sub-cutaneous tissues; and thorns and other sharp particles can cause short-lasting acute inflammatory reactions. The surgeon may introduce damaging particulate matter at operation. The best known example is talc granulomas forming in the abdominal cavity due to talc coming from the surgeon's gloves.

Once inside the body a particle may dissolve or be taken up by phagocytic cells without causing damage. Normally some inflammatory response would be expected as discussed in Chapter 5. In addition extrinsic particles might form a nucleus for epitaxial growth of intrinsic crystals.

Since they are relatively well protected, joints are not prone to disease caused by extrinsic particles. Plant thorns occasionally penetrate the capsule and cause arthritis. Injection of crystalline materials into joints and other cavities can also cause problems. The crystalline cortico-steroids used to treat inflammatory joint disease sometimes cause a transient flare up of the synovial reaction before the therapeutic benefit is felt. This may well be due to inflammatory reaction to the crystalline particles in the preparation.

3.4   Particle-induced diseases of joints

We have discussed how crystals may grow in connective tissue, or be introduced to the body from outside. Particles may also appear in joints by being shed from the bone or cartilage components themselves.

Once inside the joint, they may interact with other pathological processes to produce damage. Large deposits may damage the structures mechanically, whereas individual crystals can cause inflammatory reactions (Fig. 3.24). Thus particles are associated with both acute inflammatory diseases and chronic destructive changes.

In later chapters these conditions will be described according to the scheme laid out in Fig. 3.25. The metabolic background leading to the accumulation of high concentrations of the basic metabolite will be described further. From what is known about the formation of crystals *in vitro*, including both nucleation and crystal growth, we will then discuss the conditions determining precipitation *in vivo*. The structure and form of the crystals, and the pathology of the tissues in which they are found, will be covered, including some discussion on the pathogenesis of tissue damage. Finally the clinical associations will be reviewed in an attempt to see whether our basic knowledge of the condition accords with clinically recorded facts. As in all branches of medicine, we found that this approach can produce more questions than it answers. Indeed, we have tried to avoid any assumptions, even the one that postulates a relationship between crystals and disease!

Several other rheumatic diseases are associated with the deposition of crystals in the joints, periarticular tissue or elsewhere. Deposits outside synovial joints almost invariably contain hydroxyapatite and the calcinosis of scleroderma, dermatomyositis and other connective tissue diseases are examples of apatite deposition, and are discussed further in Chapter 8.

Many of the principles outlined in this book are relevant to the diseases of other organs, such as gall stones and silicosis. These diseases can profitably be approached using the same scheme. Although this will not be attempted in this book, which is concerned primarily with rheumatic diseases, some

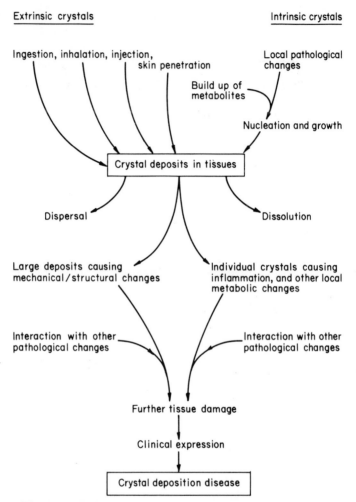

Fig. 3.24 Pathways involved in a crystal deposition disease.

useful lessons will be borrowed from other regional disciplines in a later chapter (Chapter 10).

A crystal deposition disease is therefore seen as a pathological entity which can be approached in a scientific manner and which can only be understood if basic understanding of the formation and dissolution of the particles is apparent, and where the ability of the particles to cause disease and interact with other pathological processes is known. Many diseases affecting many organs are associated with the presence of crystals, and one hopes that better understanding will lead to better treatment of the group as a whole.

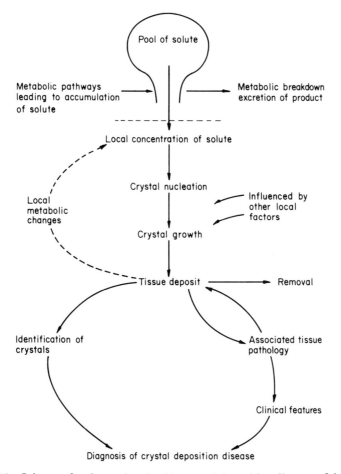

Fig. 3.25   Scheme of pathways involved in crystal deposition diseases of the joints.

3.5   Summary

In crystals, molecules are in their lowest energy arrangement, regularly packed to achieve the highest density and strong intermolecular bonding. The characteristic properties of crystals stem from this high density and the difficulty of disrupting such a regular array.

Crystals will dissolve unless supersaturation is maintained, at least locally. For ionic compounds saturation is defined by the solubility product. This may be increased if one or other of the ions is chelated (bound) to other components of the solution. Spontaneous nucleation does not normally occur until quite high supersaturations but the concentration at which this happens can be reduced by the presence of other solid particles on whose surface nucleation can occur. Often there is a constant orientation between the structures of the nucleating particle and the growing crystal (epitaxy). Once nucleation has occurred crystals grow at a rate which is dependent on supersaturation and may be imperceptible at low supersaturation (below 50%). In the presence of crystal poisons the nucleation and growth rates can drop to zero.

Crystals may grow in joint tissues, or be introduced from outside (intrinsic and extrinsic crystals). Removal will depend on dissolution (proportional to saturation) and cellular mechanisms such as phagocytosis. Any particle found within a joint is a potential pathogen.

Further reading

BRICE, J.C. (1973) *The growth of crystals from liquids*, North Holland, Amsterdam.

FARADAY DISCUSSIONS OF THE CHEMICAL SOCIETY (1976) 'Precipitation' No.6.

PHILLIPS, F.C. (1971) *An Introduction to Crystallography*, 4th edn, Oliver and Boyd, Edinburgh.

ZIPKINS, I. (ed) (1973) *Biological Mineralisation*, Wiley, New York.

Text references

BOOTH, A.H. and BUCKLEY, H.E. (1955) *Canad. J. Chem.* **33**, 1155, 1162.

BRAGG, W.L. (1961) The Rutherford Memorial Lecture. *Proc. Roy. Soc.* **262**, 145.

BUCKLEY, H.E. (1951) *Crystal Growth*, Wiley, New York.

GILMER, G.H. (1980) Computer Models of crystal growth. *Science* **208**, 355.

KATZ, W.A. and SCHUBERT, M. (1970) The Interaction of Monosodium Urate with Connective Tissue Components. *J. Clin. Inves.* **49**, 1783.

MOORE, W.J. (1972) *Physical Chemistry*, 5th edn, Longmans, London.

MULLIN, J.W. and GASKA, C. (1969) Growth and dissolution of potassium sulphate. *Canad. J. Chem. Eng.* **47**, 483.

NANCOLLAS, G.H. (1979) The growth of crystals in solution. *Adv. Colloid Interface Sci.* **10**, 215.

PERRICONE, E. and BRANDT, K.D. (1978) Enhancement of Urate Solubility by Connective Tissue I. *Arth. Rheum.* **21**, 453.

PERRICONE, E. and BRANDT, K.D. (1979) Enhancement of Urate Solubility by Connective Tissue II. *Ann. Rheum. Dis.* **38**, 467.

RINAUDO, C. and BOISTELLE, R. (1982) Theoretical and experimental growth morphologies of sodium urate crystals. *J. Crystal Growth* **57**, 432.

TROOST, S. (1972) *J. Crystal Growth* **13/14**, 449.

UHLMANN, D.R. (1971) In *Elasticity, plasticity and structure of matter* (ed. R. Houwink), 3rd edn, Cambridge University Press, pp. 292.

WILCOX, W.R., KHALAF, A., WEINBERGER, A., KIPPEN, I. and KLINENBERG, J.R. (1972) Solubility of Uric acid and monosidum urate. *Med. Biol. Eng.* **10**, 522.

Chapter 4

THE IDENTIFICATION
OF CRYSTALS

4.1 Introduction

The history of the crystal related arthropathies has followed the development of new techniques for identifying crystals and their precursors (Table 4.1).

Table 4.1 Technical advances aiding development of understanding in crystal deposition diseases

(1) Documentation of clinical details
(2) Chemical analysis of calculi and tophi
(3) Animal models of inflammation
(4) Radiographic imaging
(5) Powder diffraction of crystalline deposits
(6) Aspiration of joints
(7) Polarized light microscopy of synovial fluids
(8) Analytical electron microscopy and other technical advances in analysis

Sir Alfred Baring Garrod (1876, Fig. 1.1) used the murexide test, and the now famous 'string test' to identify raised uric acid levels in the blood of gouty patients. Subsequently McCarty and Hollander (1961) used polarized light microscopy to help identify the crystals themselves in joint fluids. More recently, analytical electron microscopy, combined with diffraction studies, have aided the identification of other salts, such as hydroxyapatite. Joint radiography has also played an important part, allowing the routine identification of articular calcification (chondrocalcinosis).

The story is surely not complete, and further advances can be expected as analytical techniques develop. For this reason we believe that this book, which endeavours to cover the scientific background to crystal formation and identification, is timely.

4.2   Identification of unknown chemicals

Despite the great range of powerful instrumental techniques available to a modern chemist, the identification of an unknown compound of unknown origin is not an easy task. The job becomes easier if the source of the material allows one to deduce something about its possible composition. Thus the identification of particulate materials from the body is often quite simple because the range of possibilities is limited and the properties of most of this group will have been tabulated. However, most of the methods we will discuss are not completely specific so it is best to use more than one technique if the result is particularly crucial or if there is any doubt arising out of the first analysis. One special advantage of crystalline materials is that the regular arrangement of molecules present within a crystal makes it difficult for foreign molecules to be bound within it. Hence crystallization is a common method of chemical purification, and if we have a precipitate consisting of facetted, similarly shaped particles we can presume that it is essentially pure.

In the following sections we will consider methods for the identification of crystals and other particulate matter in joints and other biological specimens. We will try both to describe how to use the simple methods available in the course of clinical work, and to cover the principles of more sophisticated methods that are encountered in the research literature.

4.3   Sampling

Without good sampling techniques even the most sophisticated analytical methods are going to achieve very little. It is difficult to give many formal rules as the procedure must depend very much on the situation. Ideally the sample should be uncontaminated, protected from decomposition or change, and representative of the whole tissue. Freedom from contamination can usually be established using control samples and different methods of sample preparation. The decomposition can be studied by storing samples under various conditions and examining some fresh. The representativity can only be established by sampling at different times and sites, and often presents the greatest problem.

In the specific case of joint diseases, samples available for analysis include synovial fluid, synovium and cartilage (Table 4.2).

4.3.1.   Synovial fluid

Needle aspiration of fluid from synovial joints is a relatively safe, simple procedure. Provided that precautions are taken against infecting the joint, there is no great risk; and with a knowledge of the anatomy involved most

Table 4.2

Available tissue	Usual means of sampling
Synovial fluid	Needle aspiration
Synovial membrane	Blind biopsy or arthroscopy
Articular cartilage	Open biopsy
Periarticular soft tissues	Direct biopsy or needle aspiration

joints can be tapped easily and without much discomfort. (The techniques of joint aspiration are described elsewhere, see References and Further reading).

Crystal-induced synovitis is associated with the presence of crystals in the synovial fluid, as well as in the synovial membrane. As the fluid is easily accessible, joint aspiration and its examination for crystals or alternative causes of synovitis (such as infection) is an important diagnostic procedure. If possible, all cases of acute or undiagnosed arthritis should have fluid aspirated to help confirm or exclude the presence of infectious organisms or crystals. The skin is cleaned, and using a 'no-touch' technique to avoid contamination, a needle is passed through the skin and subcutaneous tissue, and into the joint cavity. Fluid is aspirated into a clean, sterile syringe, and the needle withdrawn. The fluid can then be examined.

Normal synovial fluid is a clear, viscous yellowish fluid containing small numbers of synovial lining cells and white blood cells (about 200 cells/μl). Inflammation of the joint decreases the viscosity as the polymeric hyaluronate is broken down, and the numbers of white blood cells present increases; polymorphonuclear leucocytes are characteristic of acute inflammation and rheumatoid arthritis, whereas mononuclear cells dominate in more chronic inflammatory diseases. (Figs 4.1 and 4.2 show the total and differential white cell counts of fluid collected in a variety of different rheumatic diseases.) The cause of joint inflammation may also be apparent from synovial fluid analysis: in addition to crystals or infective organisms, immunological stimuli such as immune complexes may be present, and the products of immunological reactions including split products of complement may be found. Synovioanalysis should therefore include gross inspection of the fluid, a total and differential white cell count, culture for infective organisms, and in some cases special biochemical or immunological tests to help define the nature of the joint pathology. As shown in Fig. 4.3, fluid should be collected in a tube containing anticoagulant for the cell count, in a sterile container for culture, and in plain tubes for crystal analysis and other tests.

A drop of uncentrifuged fresh fluid can be placed on a clean microscope

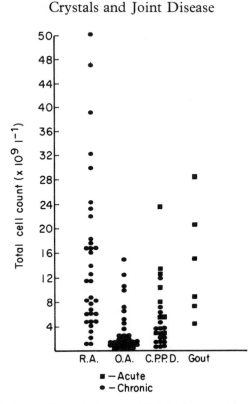

Fig. 4.1   Total white cell counts in synovial fluid from patients with rheumatoid arthritis (RA), osteoarthritis (PA), pyrophosphate arthropathy (CPPD) and gout.

slide, and examined for crystals, as outlined in Appendix I. Anticoagulants and preservatives should be avoided, as they can result in birefringent artefacts on light microscopy. In most examinations for routine diagnostic purposes, further processing of the synovial fluid is not undertaken. Three other techniques are available to aid identification of particles: (1) the fluid can be centrifuged, and the cell pellet processed for histological and electron microscope examination like a tissue sample; (2) the fluid can be processed in a cytospin centrifuge, or filtered, in an attempt to concentrate or extract the particles, although fibrin and collagen strands and other debris in the samples often make this difficult; (3) ferrography has recently been used to extract particles. The ferrograph is an instrument which magnetizes particles by exposing the sample to magnetic ions which adsorb to the particles and then extracts them by passing the fluid over a magnet. The magnetic surface can be covered with material on which to collect the particles, such as a filter paper or glass slide. This technique is being evaluated at present (Mears *et al.*, 1978).

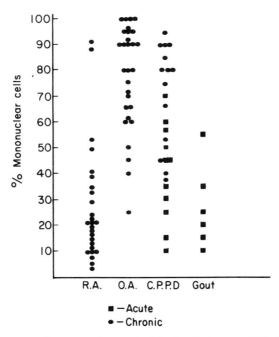

Fig. 4.2   Percentage of mononuclear white blood cells in synovial fluids of patients mentioned in Fig. 4.1.

Although it is easy to obtain, several errors can arise when analysing synovial fluid. The sampling error can occur, crystals not being present in the drop obtained from a particular pocket of the joint tapped, contaminants can enter, and look like crystals; and crystals could theoretically grow or dissolve after the fluid has been aspirated. On the basis of the known growth characteristics of most of the crystals involved in joint disease, significant errors from *in vitro* growth or dissolution seem unlikely (see Chapters 6 to 8), and this accords with clinical experience. However, brushite crystals (dicalcium phosphate dihydrate) can form in fluids on the bench, especially if they are left overnight, uncovered (so that CO_2 escapes, altering the pH). Collection of samples under oil avoids this problem.

4.3.2   *Biopsy samples*

Joint tissue can be obtained by percutaneous biopsy, by arthroscopic examination, or by operation. Percutaneous biopsy is relatively simple, but the desired sample is not always obtained, and if there are patchy changes in the joint, there is no way of knowing if an involved area has been sampled or not. Arthroscopy has the advantage of allowing biopsy under direct vision, and also allows the operator to examine the surface of the cartilage and

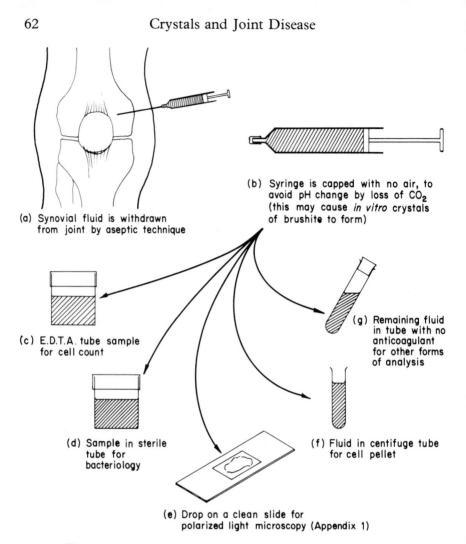

Fig. 4.3   Collection of synovial fluid, and subsequent analysis.

synovium. Large deposits of crystals are often visible to the naked eye during arthroscopy (Fig. 4.4–see colour Plate 1). Open biopsy at a formal operation is often the only way to get satisfactory samples, particularly in joints other than the knee, when the other procedures are more difficult.

Tissue samples can be examined histologically to assess the cellular reactions and tissue damage, and for the presence of crystals. In some cases culture for organisms (especially T.B.), and special biochemical and staining techniques to identify particular substances may be useful. For crystal analysis the samples can be processed for light microscopy or electron microscopy. Neutral buffers and stains should be used to avoid dissolving crystals, but no other special precautions need be taken during processing. However, if thin sections are cut, the deposits may fall out of the sliver of tissue, and many published photomicrographs of crystal deposits are in fact of 'holes' where a crystal may have been (Fig. 4.5). Thicker sections are particularly important in electron microscopy, although crystals are retained at the expense of poorer quality pictures, unless a high voltage machine is available (*vide infra*).

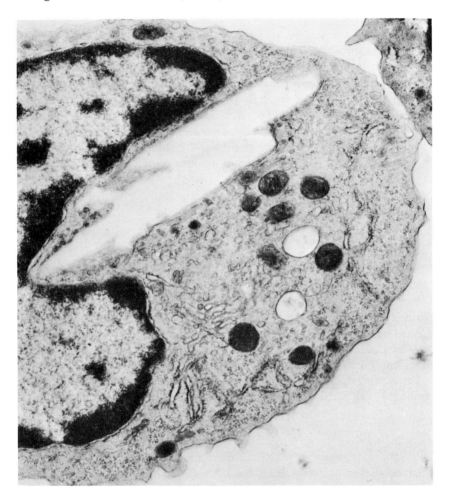

Fig. 4.5 Transmission electron micrograph of a phagocytic cell from synovial fluid in a case of gout. The thin cut has allowed the crystal to fall out of the section, leaving a 'cleft' in the cell where the crystal was. Note the close proximity of the crystal to the cell nucleus ($\times 22\,500$).

The main problem in tissue sections is finding any particles within the minute sample that reaches the microscope. Various attempts have been made to try and improve the 'yield', and aid correlations between light and electron microscopy. The recently described 'micro-incineration' method looks useful, and allows rapid processing by light microscopy of large areas to identify regions of interest for the scanning transmission electron microscope.

4.3.3   Tophi, stones and crystal deposits

The techniques described so far are mainly used to identify single small particles within tissue or fluid samples. If tophi, stones or other microscopic aggregates of crystals are present, identification is much easier. Visible deposits can be extracted from tissue surgically, washed in water, dried, and ground up for analysis by light microscopy, or by the more precise techniques such as infra-red spectroscopy and X-ray diffraction. Washing helps to remove organic matter, and solutes such as NaCl which would otherwise crystallize on drying; the relative insolubility of the compounds in question means that no significant amount of material is lost.

Organic solvents and enzymes can be used to aid extraction of organic material from aggregates, although problems could arise from inducing a different crystal phase with these manipulations.

4.3.4   Analysing joint specimens

Material obtained from in and around the joints can be processed along the lines of the scheme shown in Fig. 4.6. All specimens can be examined in a polarized light microscope. Macroscopic deposits can be identified by X-ray powder diffraction or infra-red spectroscopy. The identification of small individual particles may be aided by analytical electron microscopy. Standardization of the techniques used against known pure samples of the likely compounds involved is helpful.

In the remainder of this chapter, the principles of, and information obtainable from, these different analytical instruments will be outlined.

4.4 Light microscopy

The light microscope is the only instrument available to most clinicians for the identification of crystals. It has both advantages and disadvantages.

The instrument is simple, cheap and easy to use. On the other hand it lacks the ability of most of the other techniques described to positively identify an unknown crystal. In the section that follows an outline of the principles behind polarized light microscopy is presented; Appendix I contains a guide to the technique of examining synovial fluids.

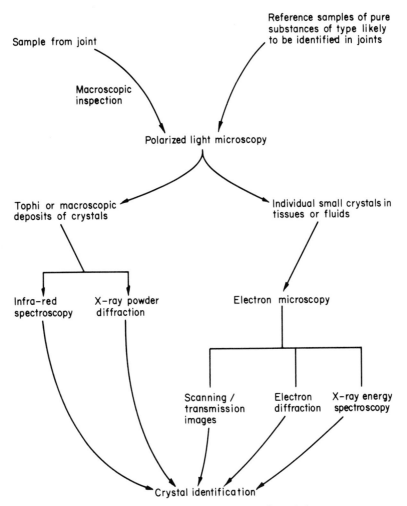

Fig. 4.6 Means of analysing samples from joints.

4.4.1 *The optical activity of crystals*

The interactions of light with crystals stem from two fundamental processes, (1) the absorption of light by molecules of a suitable structure, and (2) the dependence of the velocity of transmission of light on the density of the medium (refractive index).

Absorption in the visible region is common in large aromatic organic compounds and in compounds of the transition metals such as iron and copper. Since the absorption will not usually be uniform across the visible wavelengths these compounds are coloured. Such crystals can often be identified by colour alone, although this may also result from impurities within the crystal.

The refractive index of a medium is given by the ratio of the velocity of light in a vacuum to that in the medium. In homogeneous materials such as glasses and liquids the velocity of light decreases as the density increases, and thus the refractive index increases with density. The fact that light changes direction when it crosses a refractive index boundary means that amorphous or crystalline particles will be visible when immersed in a liquid of different refractive index. This difference can be made more apparent by using phase-contrast illumination to give an intensity of colour contrast. If the particles are very small or have a very rough surface they will appear dark as the light is scattered at the surface. By changing the refractive index of the liquid so that it almost matches that of the particle it is possible to measure the particle's refractive index. Given a perfect refractive index match, of course, the particle vanishes completely. A series of liquids for refractive index measurement is available (Cargille liquids) though these are, of course, only useful for insoluble particles. Refractive indices of liquids can readily be measured using an Abbé refractometer.

The situation with crystals is more complicated because their structure is not the same in all directions. As a consequence the velocity of light will depend on the direction of propagation, i.e. they will be birefringent. Light propagates as an electrical field oscillating perpendicular to its direction of travel. Normally the oscillation itself has no particular direction but in a crystal waves with oscillations in different directions can travel at different velocities. A beam of light travelling towards us can be considered as having only two directions of oscillation: vertical and horizontal; all other directions can be treated as a sum of these two. We can remove one of these two by passing the light through a polarizing film which selectively absorbs one of the two directions (Fig. 4.7). If this light then passes through a crystal we find that the refractive index depends on the direction of oscillation in relation to the axes of the crystal, this is the principle behind the use of polarized light microscopy to identify crystals.

The behaviour of the crystal is complicated, but its refractive indices can be treated in terms of three principal directions defining the main axes of an ellipsoid, the indicatrix. These axes are frequently simply related to the actual shape of the crystal and to its internal symmetry (Fig. 4.8). If the crystal is cubic, like diamond or common salt, the axes are all of equal length, the 'ellipsoid' is a sphere and the crystal has the same non-directional optical properties as glass. If two of the three axes are equal the crystal is uniaxial (hexagonal, tetragonal and trigonal crystals). If such a crystal happens to be flat (lozenge shaped) so that it sits on the microscope slide with its optic axis upwards then light passing up through the microscope will experience no special effect as the refractive index is independent of the direction of oscillation for a wave propagating along the optic axis. In all these cases if we introduce plane-polarized light into the bottom of the microscope it will pass unrotated through the particle.

In a standard polarized light microscope two plane polarizers are used, one below and one above the sample stage. If the upper polarizer is orientated at right angles to the lower one, the eye sees a dark background. any birefringent crystals will then appear light against this dark background (Fig. 4.8(a)).

If the crystal is uniaxial and needle-shaped so that its optic axis is in the plane of the slide, light passing through will be split into two waves, one parallel and one perpendicular to the optic axis, travelling at different

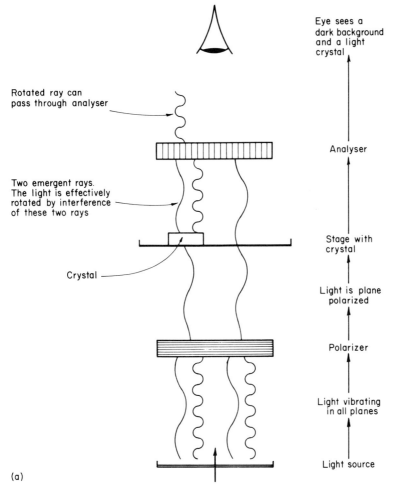

Eye sees a
dark background
and a light
crystal

Rotated ray can
pass through analyser

Analyser

Two emergent rays.
The light is effectively
rotated by interference
of these two rays

Stage with
crystal

Crystal

Light is plane
polarized

Polarizer

Light vibrating
in all planes

Light source

(a)

Fig. 4.7 (a) Simplified diagram of the polarized light microscope. (b) See overleaf.

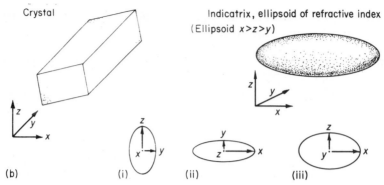

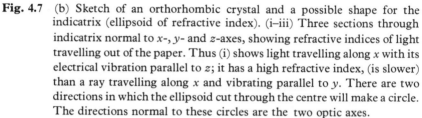

Fig. 4.7 (b) Sketch of an orthorhombic crystal and a possible shape for the indicatrix (ellipsoid of refractive index). (i–iii) Three sections through indicatrix normal to x-, y- and z-axes, showing refractive indices of light travelling out of the paper. Thus (i) shows light travelling along x with its electrical vibration parallel to z; it has a high refractive index, (is slower) than a ray travelling along x and vibrating parallel to y. There are two directions in which the ellipsoid cut through the centre will make a circle. The directions normal to these circles are the two optic axes.

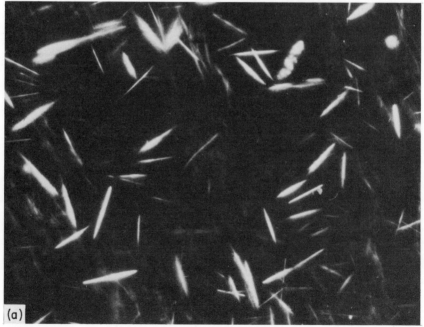

Fig. 4.8 (a)

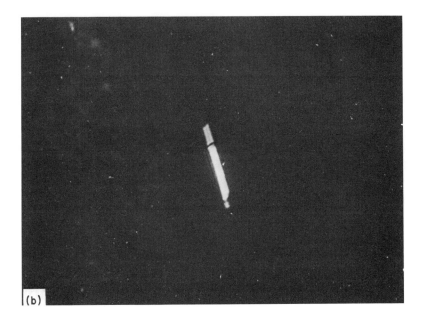

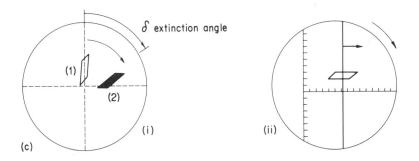

Fig. 4.8 (a) Synovial fluid urate crystals seen in polarized light microscope, between crossed polars. Note the light, needle-shaped crystals on the dark background (× 570). (b) A single pyrophosphate crystal viewed between crossed polars (× 600). (c) Estimating the extinction angle and dimensions of a crystal in the polarized light microscope. (i) Take out analyser, align crystal long edge parallel to cross wire of eyepiece (and to polarizer). Insert analyser and rotate crystal to nearest point of extinction. Rotation is extinction angle. (ii) Rotate stage to align crystal with a grating or moving hair in eyepiece. Calibrate grating against a steel mm rule.

velocities. This means that when they emerge from the crystal the plane of polarization will have been effectively rotated. This effect will be maximum when the optic axis, usually also the needle axis, is at 45° to each of the polarizer directions and zero when it is parallel to one or the other. The angle, if any, between the optic axis and the crystal axes is the extinction angle which can readily be measured (Fig. 4.8(c)). Similar effects are seen with biaxial crystals where all three axes are different but the crystal will go bright and dark when the microscope stage is rotated whatever face of the crystal is uppermost.

Thus brightness between crossed polars, birefringence, is a sign of crystalline material, but it is not exclusive to crystals. Care needs to be taken to avoid dust, hairs and fibres and weak effects due to reflections from the surface of other amorphous particles. Collagen particles and some proteins will also appear crystalline, as they have unidirectional order but will lack the regular outline of a crystal. Single crystals will go from bright to dark every 45° as the stage is rotated, but clusters of very small crystals may just show an overall but shifting brightness. If the crystals are much smaller than the wavelength of light and lack any overall group order they will not appear bright.

It is possible to go some way towards quantifying the birefringence. Small or weakly birefringent crystals will simply appear white between crossed polars but as the thickness increases, so does the difference in travelling time for the fast and slow rays. As this path length difference becomes a significant fraction of a wavelength, the crystals appear coloured as the shorter wavelengths are rotated by 180° and so extinguish again (Fig. 4.9).

The path length difference of light rays in a crystal is given by

$$R = \Delta n \cdot t$$

where Δn is the refractive index difference and t is the thickness. The rotation is given by $\phi = 2\pi R/\lambda$ in radians or $360R/\lambda$ in degrees. When the crystal is very thick the colours fade to white again. It is possible to convert colours to R values. If the crystal appears white a mica or quartz crystal can be introduced at 45° to the polars such that the whole field appears violet (first-order red compensator, $R = 575$ nm). The test crystal will now cause the pink colour to shift towards yellow or blue to an extent depending on the sign and magnitude of R. If the crystal optical axis is parallel to the axis of the plate a *shift to blue corresponds to positive birefringence Δn, to yellow negative Δn* (Fig. 4.10). If t can be measured for the crystals, Δn can be estimated in this way, although it is usually sufficient just to characterize birefringence as weak or strong, positive or negative. A thorough series of measurements requires a microscope stage on which a single crystal can be rotated to any angle, and a variety of adjustable compensators, as well as the ability to measure thickness.

R in mμ	Increasing thickness of wedge	Wedge at 45°. Crossed Polars	R in λ. Colours extinguished.	Resulting interference colours. Crossed Polars	Order.	Colours with Parallel Polars
0				Black		Bright White
				Iron Grey		White
100				Lavender Grey		Yellowish White
				Greyish Blue		Brownish White
200				Grey	1st. Order	Brownish Yellow
				White		Light Red
300				Light Yellow		Indigo
				Yellow		Blue
400			1λ Violet			Blue Green
			1λ Blue	Orange		
500				Red		Pale Green
600			1λ Yellow	Violet		Greenish Yellow
				Indigo		Yellow
700			1λ Red	Blue	2nd. Order	Orange
				Green		Light Carmine
800			2λ Violet			Purplish Red
			2λ Blue	Yellow Green		Violet Purple
900				Yellow		Indigo
				Orange		Dark Blue
1000				Orange Red		Greenish Blue
1100			2λ Yellow	Dark Violet Red		Green
				Indigo		Pale Yellow
1200			3λ Violet	Greenish Blue	3rd. Order	Flesh Colour
1300			3λ Blue	Green		Violet
1400			2λ Red	Greenish Yellow		Greyish Blue
1500				Carmine		Green
1600			4λ Violet	Dull Purple		Dull Sea Green
			3λ Yellow	Grey Blue		Greenish Yellow
1700			4λ Blue	Bluish Green	4th. Order	Lilac
1800				Light Green		Carmine
1900				Greenish Grey		Greyish Red
2000			5λ Violet	Whitish Grey		Bluish Grey
				Flesh Red		Green
2100			3λ Red			
2200			4λ Yellow			
2300			5λ Blue			

Violet
Blue
Yellow
Red

} Shading indicates the limits between which light of the wave-lengths shown, is extinguished.

(a)

Fig. 4.9 (a) Caption overleaf.

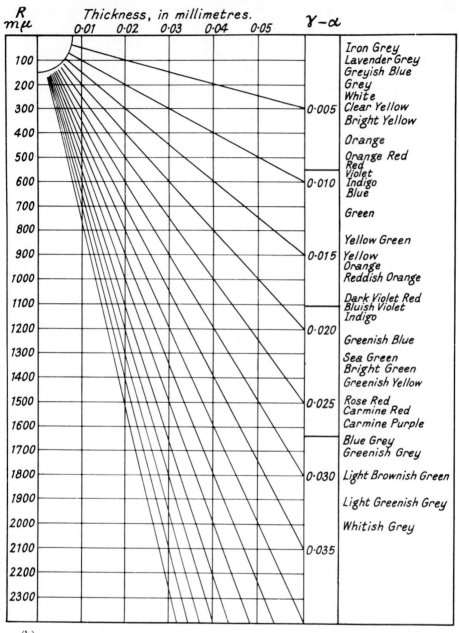

(b)

Fig. 4.9 (a) Chart of colours seen with crossed and parallel polars as a function of path length difference, R, introduced by a crystal plate. (b) Simplified version of Michel–Levy's chart for determining birefringence ($\Delta n = \gamma - \alpha$) for crystals of different thickness. (With permission of Hartshorne and Stuart, 1964.)

Fig. 4.10 Positive and negative birefringence.

(a) Small crystals or low birefringence.

Insert first-order red compensator with (slow) axis at 45° to polarizer and analyser. Rotate crystal parallel to λ (marked ↑ on compensator.).

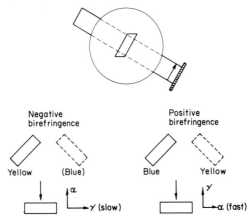

(b) Large, highly birefringent crystals.

Cause too much colour shift for the direction to be unambiguously picked out. A one-quarter λ plate is added instead, and the sign determined from the direction of shift using Fig. 4.9. (Strictly this only applies if the crystal is uniaxial with the optical axis along the needle axis, but it is used for most elongated crystals.)

Crystal size can be measured microscopically by using a filar eyepiece in which a cross-hair is moved by a calibrated drum attached to the eyepiece or more easily by an image-shearing eyepiece in which two differently coloured images are adjusted until they just touch. Thickness can be roughly measured by focusing on the microscope slide and then again on the top of the crystal, and measuring the number of gradations of the focusing wheel needed to do this. Measuring eyepieces are calibrated using an engraved slide on which 1 mm is divided into 100 units. The lens effect of a crystal whose refractive index differs from the surrounding liquid can make measurement difficult. A better refractive index match reduces this problem. Crystal shape can be characterized by the angles between the faces and the ratio of the lengths of sides.

Crystal birefringence, size and shape are of great use in the characterization of crystals but cannot be used for a definitive identification. However, these parameters, together with photographs and information on the number and distribution of crystals, are of great help in diagnosis of a crystal deposition disease (Fig. 4.11–see colour Plate 2).

The techniques used to examine synovial fluid in the polarized light microscope are outlined in Appendix 1.

4.5 Electron microscopy

Optical microscopy is intrinsically limited by its inability to resolve objects much smaller than the wavelength of light, i.e. 500 nm. In the early 1900s attempts were made to get round this by carrying out microscopy in the ultraviolet region, but this did not really produce useful gains in resolution. An understanding of the wave nature of matter grew in the 1920s and this led to the demonstration of electron diffraction and thus to the development of the electron microscope. The potential resolution of this instrument is no longer limited by the wavelength, which is 0.004 nm for an electron accelerated by 100 kV, as occurs in a normal machine.

The transmission electron microscope is very similar in construction to an optical microscope. The electron gun at the top acts as a lamp, electrical or magnetic condenser lenses focus the beam on to the sample and objective and projector lenses produce an image on a fluorescent screen at the base of the microscope. Image contrast comes from both absorption and diffraction effects as in a normal microscope but electrons interact strongly with matter, especially high atomic weight elements, so the sample must be very thin to allow the beam through. Similarly the column must be evacuated as electrons cannot penetrate far in air. A large amount of information is available from transmitted, diffracted or emitted energy resulting from a focused electron beam striking the specimen (Fig. 4.12).

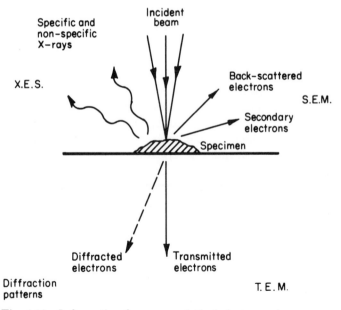

Fig. 4.12 Information from an analytical electron microscope.

4.5.1   Transmission electron microscopy

The transmission microscope offers the greatest magnification and resolution for suitable samples, but sample preparation is a major problem. The sample must be very thin in order to be transparent to electrons but must also be very robust to withstand the very high radiation levels. The radiation intensity in an electron microscope has been compared to that experienced within ~ 100 yards of an exploding nuclear bomb. For biological samples it is usual to cut thin sections of about 1000 Å from frozen or epoxy embedded and fixed tissue. Contrast may be enhanced by use of heavy metal selective stains. In this processing changes may be induced in the sample and crystals may be dissolved or dislodged so that only crystal-shaped holes are seen (Fig. 4.5). Whilst resolutions of 1 to 2 Å are claimed by microscope manufacturers, this really only applies to high melting-point crystalline materials; in organic materials the limiting resolution is about 20 Å. The technique can be very informative and given a good recipe for sample preparation it is reasonable to expect good pictures within the first few attempts. Developing new methods for sample preparation is a major exercise and great care must be taken to avoid artefacts.

4.5.2   Electron diffraction

By adjusting the electron lenses to produce a parallel rather than a converging beam it is possible to get crystal diffraction patterns in an electron microscope. Pinhole apertures in the beam allow small selected areas of the sample to be studied in this way. The resulting patterns are not as easy to interpret as are X-ray diffraction patterns as the effective electron wavelength is much smaller so many high-order reflections are seen and also crystals tend to lie on one or two faces only, so that all reflections cannot be seen. Sample stages can be tilted to improve this. Since the destructive power of the beam will often disrupt internal crystal structure before changing the external appearance it is usually advisable to search for crystals at low magnifications with a low beam intensity and then record the diffraction pattern before going to high magnification. The 'foamy' pattern seen with many crystals under electron microscopy, such as pyrophosphates from joints, is due to beam damage, and limits the usefulness of diffraction and the other analytical techniques described below (Fig. 4.13).

4.5.3   Scanning electron microscopy

In the scanning microscope the electron beam is focused to a fine spot on the sample surface and some of the scattered electrons are picked up by a detector mounted at an angle to the beam. The picture is formed by scanning the beam over the sample surface in a TV type of raster pattern and using the

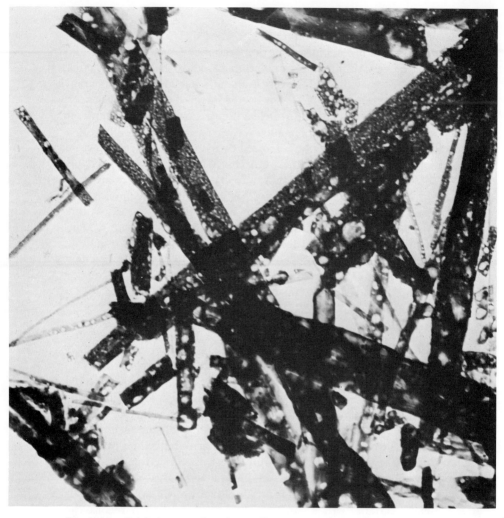

Fig. 4.13 Calcium pyrophosphate dihydrate crystals viewed in transmission electron microscope. Note the 'foamy' appearance of the crystals due to beam damage (× 14 000).

detector output to form the image on a TV screen. The sample may be mounted at an angle to the beam or perpendicular to it, the image contrast arises mainly from the topography of the sample surface affecting the number of electrons reflected towards the detector.

Samples are very easy to prepare, usually they are cut or broken so as to produce a rough surface whose features will betray the internal structure. They are cemented to a metal stud and sputtered with gold to produce a conducting surface, a process that takes only a few minutes. The sample must, of course, be stable in a vacuum. SEM pictures are characteristic for their clarity and great depth of focus. Magnification can be readily changed from a few times to about × 10 000. Resolution is normally limited to about 20 Å dependent on the hardness of the sample.

Whilst the pictures produced are very striking and make excellent illustrations, they are not necessarily very informative (Fig. 4.14). Accordingly, it is wise to give careful thought to whether any specific features should be identifiable before starting. In samples with a relatively smooth surface the appearance is similar to that of a landscape viewed from above when the sun is low. Accordingly it can be difficult to tell the hills from the valleys unless some feature is identified which is unmistakably a hole or a peak and which can give an indication of the direction of the 'sun'. It is important to keep track of the orientation of pictures through the process of developing. Also comparisons should be made at low magnification between fresh areas of sample and those possibly damaged by exposure to the beam at high magnification.

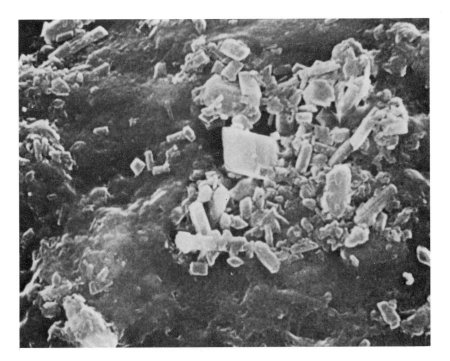

Fig. 4.14   Scanning electron micrograph of pyrophosphate crystals on the surface of joint cartilage (× 6000).

4.5.4   Electron analytical techniques

In recent years a wide range of analytical facilities has been developed for use in conjunction with electron microscopes to the extent where scanning microscopes tend to resemble Christmas trees. It is possible to derive chemical information on a sample from the emitted X-rays, secondary and scattered electrons, and atoms and ions released on bombardment of the sample with the electron beam (Fig. 4.12). The method most directly applicable to crystal identification is the analysis of the X-rays produced by the impact of the electron beam. The X-rays are analysed either by energy-sensitive or wavelength-sensitive detectors. The two are somewhat complementary in that the wavelength-sensitive method allows analysis of lighter atoms down to carbon whilst the energy-dispersive method (EDAX) is more sensitive to low concentrations. In either case the electron beam is directed at the particle of interest and its X-ray spectrum is collected. From this the concentrations of the various elements in the particle can be determined (Fig. 4.15). Great precision is not possible as the ability of the electrons to penetrate the sample, and of the X-rays to emerge, depends on the composition and topography of the sample. Typically the electrons penetrate about 100 Å into the sample but this depth can be adjusted by varying the angle of the sample to the beam. Also it is not always possible to direct the beam at a single uncontaminated particle. Given local high concentrations of a particular element it is also possible to produce a picture showing local concentrations of the element as bright areas on a dark background.

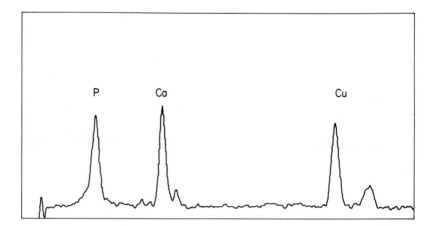

Fig. 4.15   X-ray energy spectrum of a calcium pyrophosphate dihydrate crystal, showing the calcium and phosphorous peaks, in a ratio of about 0.86, Ca:P. (The copper peak is from the microscope grid.)

Since the development of the high-intensity field emission gun, *scanning transmission electron microscopes* (STEM) have been developed which can take somewhat thicker samples than a normal transmission microscope, and can have analytical systems attached, but otherwise resemble a normal transmission microscope in their method of use. The field-emission gun produces a very high-intensity beam concentrated in a very small area, the typical diameter of the spot reaching the sample being about 50 Å. This beam is made to scan the sample as in scanning microscopes, and the transmitted image is of high quality even when thicker than normal sections are used, because of the high beam intensity. The instrument also has advantages for X-ray emission analysis; the 'spot' can be positioned precisely on any small area of the section, and the high-energy input causes enough X-ray emission from minute particles to allow elemental analysis. STEM was used in this way to make the original identification of hydroxyapatite crystals in joint fluid and tissues (Dieppe *et al.*, 1976).

4.6 Infra-red spectroscopy

All molecules are in constant vibrational motion whose intensity increases with temperature. The vibrations have characteristic frequencies which are dependent on the types of bonds in the molecule. The frequency of vibration of any bond type is determined primarily by the bond strength and the weights of the two atoms joined in the same way as is the vibrational frequency of two weights joined by a spring. Thus if the weights or atoms are heavier the vibration is slower.

The characteristic frequencies of many common bonds are in the range 10^{13}–10^{14} Hz which is the frequency of infra-red radiation of wavelength 3–30 μm. This means that in most molecules the bonds can absorb this infra-red radiation and convert the energy into vibrational motion. Clearly if this were the whole story one could measure these absorptions, determine all the bond types in the molecule and so find its structure. One cannot always do this because interactions between the various bonds on the molecule and between neighbouring molecules make the situation more complicated. However, characteristic frequency ranges are tabulated, and a few examples are shown in Figs 4.16 and 4.17. Many peaks in complex infra-red spectra can be assigned from standard tables. For example, the $C = O$ peak at about 1700 cm^{-1} is particularly useful in this respect as it is very strong and in a relatively clear part of the spectrum. It can be used to distinguish between ketones, acids and esters.

In the region from 8 to 15 μm the spectra is complicated by interactions within the molecule so that the pattern is unpredictable but very characteristic, this is therefore known as the 'fingerprint' region and is most useful to establishing the identity between an unknown and a known sample. Libraries of infra-red spectra are stored on file cards for use in identification in this way. Infra-red is very useful in this qualitative way, but it is a much poorer quantitative technique than visible or ultraviolet spectroscopy and must only be used with great caution for quantitative analysis.

A general purpose spectrometer uses a bar or filament heated to a dull red as a source. Two beams are focused by a system of mirrors through a sample and a reference, which is usually just air. The two beams are then passed alternately through a rock salt prism or on to a diffraction grating to separate the set frequency and finally focused on to a detector. The detector records the difference in signal between the reference beam and that which has passed through the sample, and gives a readout to a chart recorder. Such instruments cost from about £1000 upwards and are an essential part of almost any chemical laboratory.

If the sample is a liquid, it is usually spread thinly between polished single-crystal plates of sodium chloride and put in one beam. Solid samples can be ground in a pestle and mortar and mixed with a pure paraffin oil such as Nujol then poured between salt plates. For good resolution only a little sample must be used and the sandwich of the mull between the plates should appear transparent. Alternatively, the powder can be ground with solid potassium bromide and the mixture compressed to make a thin clear disc. The potassium bromide should be carefully dried and kept dry or a water spectrum will be superimposed on the sample. The spectrum is presented as transmittance versus frequency, often measured in wavenumbers (cm^{-1}) which is the number of wavelengths per centimetre (Fig. 4.16). The absorption peaks stretch down from the top of the plot and the scale is essentially logarithmic in concentration so a doubling in peak height corresponds to a more than double concentration and at high concentrations peaks tend to round off and detail is lost.

Fourier transform infra-red spectroscopy (FTIR) has recently become available with much greater sensitivity at low concentrations. Laser raman spectroscopy is a closely related technique where the vibrational spectrum is seen as a perturbation on the single sharp peak of visible radiation from a laser.

Infra-red spectroscopy in the low frequency range is very useful in distinguishing different calcium phosphates that may be deposited in joints (Fig. 4.17).

4.7 X-ray diffraction

X-rays are electromagnetic radiation with wavelengths in the range from about 0.1–1 nm compared to 400–700 nm for light; their energy is higher in inverse proportion to their wavelength. They are produced by accelerating

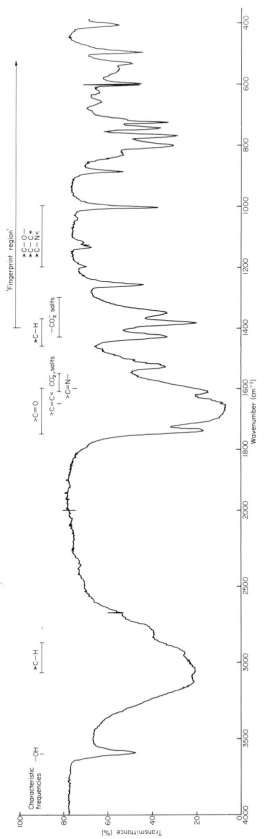

Fig. 4.16 Infra-red spectrum of monosodium urate monohydrate showing the areas of absorption caused by different bonds.

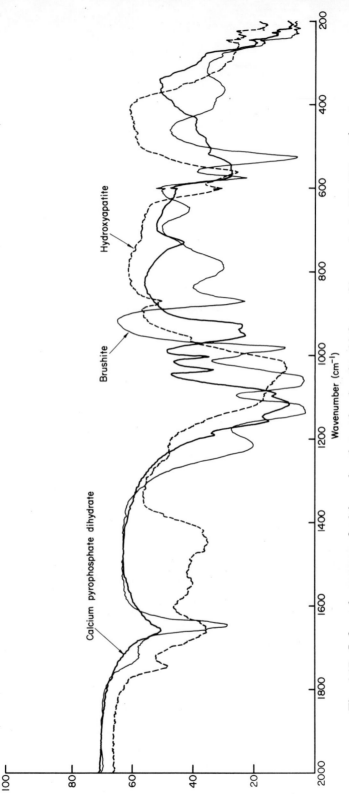

Fig. 4.17 Infra-red spectrum of calcium phosphates showing the chief area of interest in the 800–200 cm⁻¹ region.

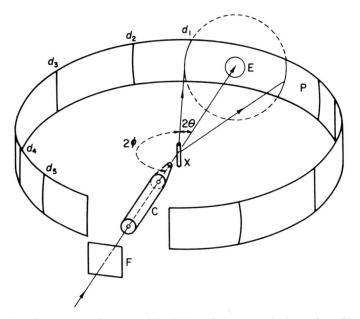

Fig. 4.18 Geometrical features of the Debye–Scherrer technique (from Klug and Alexander, 1954).

electrons along a vacuum tube with an accelerating voltage of 10–50 kV. The electrons collide with a water-cooled metal target and the ensuing electronic rearrangements result in the production of a broad range of X-rays of wavelengths down to that corresponding to the energy of the incoming electrons plus intense emissions at wavelengths characteristic of the target metal. The X-rays are emitted in all directions but are mostly absorbed inside the tube. Generally, four beams are allowed out through windows made of beryllium foil in the side of the tube.

Medical X-radiography uses 'hard' short-wavelength high-energy X-rays which are able to penetrate large distances in tissue and are absorbed to varying extents. Absorption increases with density and with the atomic weight of the elements present, hence the use of lead for shielding, and the good visualization of bone on a typical radiograph.

In crystallography the wavelengths used are those which correspond to the spacings between atoms in crystals. The X-rays are diffracted by the regular planes of atoms present in a crystal lattice in the same way as the regular array of lines on a diffraction grating diffract light.

If white light is shone on to a grating each separate wavelength will be diffracted at an angle which is characteristic of the spacing between the lines and the wavelength. As well as the colours seen in oil films on water the iridescent colours of many beetles and butterflies are partly the result of diffraction from layered structures. A crystal, being three-dimensional, is rather like a stack of diffraction gratings and this imposes additional restraints such that diffraction only occurs at an angle, 2θ, if the wavelength of the X-rays, λ, is related to the spacing between the planes, d, by Bragg's law:

$$n\lambda = 2d \sin \theta$$

where n is an integer, $n = 1$ being a 'first-order' diffraction, $n = 2$ being second order, and so on. Since $\sin \theta$ must be between 0 and 1 there are not usually more than one or two diffraction orders if λ is comparable to d. A single crystal mounted in a monochromatic X-ray beam produces a pattern of spots on a photographic film behind the crystal. The spots correspond to those directions for which Bragg's equation is satisfied.

Monochromatic X-rays are produced by filtering out with a thin metal foil all but the most intense $K\alpha$ wavelength of a target metal such as copper. In X-ray crystal-structure determination, a perfect single crystal of about 0.1–1 mm in size is mounted in the beam and then rotated whilst the diffracted spot intensities are collected either on film or by a radiation counter such as a Geiger counter. Rotating the crystal about two or three axes allows up to several thousand spot intensities to be collected. By a combination of careful guess work and computation it is then possible to determine to $\pm$ 0.001 nm (1 pm) the relative positions of the atoms in the basic repeating unit of the crystal, the unit cell. The knowledge of interatomic bond lengths and angles that derives from this is fundamental to modern chemistry.

Hydrogen atoms are not found as they have too few electrons to interact significantly with X-rays but they can be positioned in a similar way by neutron diffraction.

For routine analytical purposes the material is usually used in the form of a fine powder packed into a thin-walled glass capillary tube. The powder should be ground finely enough so that all possible crystal orientations are presented to the beam. The glass, being amorphous, gives no sharp diffraction. The sample is placed on the axis of a flat cylindrical Debye–Scherrer camera (Fig. 4.18). The X-rays travel along a diameter of the camera, strike the sample at the centre and are diffracted on to a film wrapped around the inside circumference. The sample is rotated on its axis to further randomize the crystal orientations. The developed film shows a series of circular arcs (Fig. 4.19) which are symmetrical about the beam axis. The distance between corresponding arcs can be turned directly into values of diffraction angle, θ, and so knowing the wavelength the interplanar spacings, d can be found. Ideally the pattern can be compared with that of a known compound. Alternatively the American Society for Testing Materials

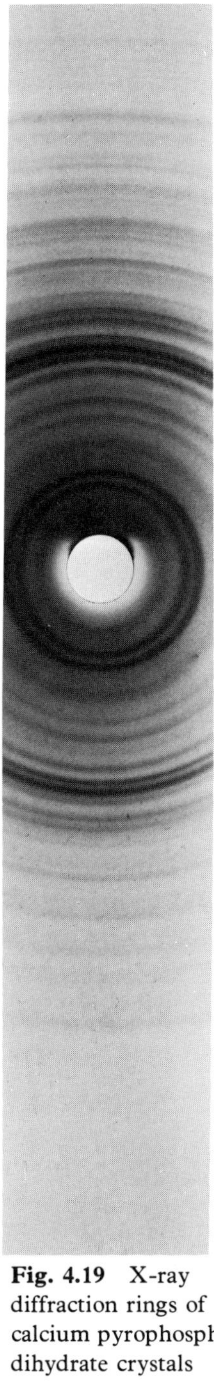

Fig. 4.19 X-ray diffraction rings of calcium pyrophosphate dihydrate crystals

publishes the ASTM index which provides *d*-spacing and intensity information on the diffraction patterns of a large number of organic and inorganic crystals. As well as tables of complete patterns, crystals are indexed according to the *d*-spacings of their three most intense lines. Thus any unknown pure powder can usually be identified from the index. Mixtures cannot really be tackled this way so microscopic examination should be used to examine whether more than one crystal type is present. It should also be remembered that crystal structures are determined by the crystallization conditions and that many compounds occur in more than one crystal form.

An X-ray generator for powder crystallography costs about £10 000, much of the money being for the very stable high-voltage power supply that is necessary for counter, as opposed to film, work. The process of taking powder patterns is straightforward but needs a considerable amount of practice before clear, sharp, unspotty arcs are produced. Organic compounds are more difficult to do because the lower atomic weight means that the intensities are lower compared to the background of scattering from the glass. When operating X-ray generators it is advisable to wear a film badge or dosimeter and to check for radiation leaks with a small counter. Modern standards require that cameras cannot be removed from the generator or opened while the beam is on. Even a brief exposure to the main beam may produce a severe burn.

Carefully done powder diffraction gives a very reliable confirmation of a compound and can be used quantitatively. It is less good for the initial identification unless the sample is pure and is not much good for detecting low concentrations of crystalline material. This limits its use in the biological sampling described in this chapter.

4.8 Indirect techniques

Several other indirect methods of identifying crystals in joints are being used or developed.

Joint radiographs may show up shadows of calcific density in or around a joint. The distribution and type of shadow may help to indicate which calcium salt is likely to be present: pyrophosphate tends to form linear shadows in cartilage, and hydroxyapatite forms spotty deposits in periarticular tissue (Chapters 7 and 8).

Special staining methods may help to identify the presence of certain salts in tissues or fluids; in addition to conventional calcium and other stains, recent work has suggested that Alizarin Red stain might, for example, help to identify synovial fluid hydroxyapatite crystals.

It may be possible to attach other substances to crystal surfaces to aid identification. Halverson and McCarty (1979) reported a method for identifying hydroxyapatite crystals using labelled diphosphonates. Similar radiolabelling techniques employing agents that bind to crystals may help in the identification of these and other crystals.

Finally, solvents may help. If urate is suspected in tissue, uricase digestion may confirm that suspicion; similarly fat solvents may remove other unidentified crystals from a tissue slice or fluid film, suggesting a lipid composition.

There are therefore a wide variety of techniques, of varying specificity and sophistication available to help in the identification of joint particles (Table 4.3).

Table 4.3 Some methods available to aid crystal identification

Polarized light microscopy
Electron microscopy
Infra-red spectroscopy
X-ray diffraction
Radiology
Staining and labelling techniques
Use of chemical solvents

4.9 Summary and conclusions

A variety of powerful analytical instruments are available to aid the identification of particles in biological specimens. However, none of the techniques used can give an exact picture of crystal structure, and fluid and tissue sampling is a 'hit-or-miss' affair, fraught with possible errors. In practice, therefore, a compromise has to be made, based on the best available samples and instruments. Accuracy can be improved by obtaining more than one specimen from the patient, and by using more than one technique of analysis, when possible.

As further technical advances are made, improved identification of the nature of particles within joints, and of their immediate environment, should increase our understanding of these diseases considerably. The future of this subject, like its history, must hinge on application of the right technique to the right sample at the right time.

Further reading

Sampling

CAWLEY, M.I.D. (1974) Arthrocentesis: techniques and indications. *Brit. J. Hosp. Med.* **11**, 74.

CROCKER, P.R., DIEPPE, P.A., TYLER, G., CHAPMAN, S.K. and WILLOUGHBY, D.A. (1976) The identification of particulate matter in biological tissues and fluids. *J. Path.* **121**, 37.

CROCKER, P.E., DOYLE, D.V. and LEVISON, D.A. (1980) A practical method for the identification of particulate and crystalline material in paraffin-embedded tissue specimens. *J. Path.* **131**, 165.

ROPES, M.W. and BAUER, W. (1953) *Synovial fluid changes in Joint Disease*, Harvard University Press, Massachussets.

Light microscopy

GATTER, R.A. (1977) Use of the Compensated Polarizing Microscope. *Clinics Rheum. Dis.* **3**, (1), 91.

HARTSHORNE, N.H. and STUART, A. (1964) *Practical Optical Crystallography*, Edward Arnold, London.

HARTSHORNE, N.H. and STUART, A. (1970) *Crystals and the Polarizing Microscope*, 4th edn, Edward Arnold, London.

WAHLSTROM, E.E. (1969) *Optical Crystallography*, 4th edn, Wiley, New York.

WOOD, E.A. (1977) *Crystals and Light*, 2nd edn, Dover Publications, New York.

Electron microscopy

CHANDLER, J.A. (1977) *X-ray microanalysis in the electron microscope*, North Holland, New York.

FRYER, J.R. (1979) *The chemical applications of transmission electron microscopy*, Academic Press, London.

HALL, T., ECHLIN, P. and KAUFFMAN, R. (eds) (1974) *Microprobe analysis as applied to cells and tissues*, Academic Press, London.

HEARLE, J.W.S., SPARROW, J.T. and CROSS, P.M. (1972) *The use of the scanning electron microscope*, Pergamon Press, Oxford.

KAY, D.H. (ed) (1965) *Techniques for electron microscopy*, 2nd edn, Blackwell, Oxford.

SWIFT, J.A. (1970) *Electron Microscopes*, Kogan Page, London.

Infra-red spectroscopy

CROSS, A.D. and JONES, R.A. (1969) *An introduction to Practical Infra-Red Spectroscopy*, 3rd edn, Butterworths, London.

X-ray diffraction

BUNN, C.W. (1961) *Chemical Crystallography*, 2nd edn, Clarendon Press, Oxford.

Text references

DIEPPE, P.A., CROCKER, P.R., HUSKISSON, E.C. and WILLOUGHBY, D.A. (1976) Apatite Deposition Disease: a new arthropathy. *Lancet* **i**, 266.

GARROD, A.B. (1876) *A treatise on gouts and rheumatic arthritis*, Longmans, London.

HALVERSON, P.B. and McCARTY, D.J. (1979) Identification of hydroxyapatite crystals in synovial fluid. *Arth. Rheum.* **22**, 389.

KLUG, H.P. and ALEXANDER, L.E. (1954) *X-ray diffraction procedures for polycrystalline and amorphous materials*, Wiley, New York.

McCARTY, D.J. and HOLLANDER, J.L. (1961) Identification of Urate Crystals in Gouty Synovial Fluid. *Ann. Intern. Med.* **54**, 542.

MEARS, D.C., HANLEY, E.N., RUTOWSKI, R. and WESTCOTT, V.C. (1978) Ferrography, its application to the study of human joint wear. *Wear* **50**, 115.

Chapter 5

CRYSTAL-INDUCED
DISEASE

5.1 Introduction

This chapter contains an outline of the ways in which crystals can cause tissue damage. Attention is focused on the mechanisms underlying acute and chronic joint damage, as these have been well investigated and are most relevant to the understanding and treatment of the conditions described in the next four chapters.

As explained in Chapter 2, crystal deposition diseases are the result of a complex sequence of events, including those which lead to deposition, and those resulting from the presence of crystals in the tissue. The mere presence of crystals is not bound to cause a disease. Post-mortem and other pathological examinations often reveal the presence of a wide variety of crystal deposits in the soft tissues, which are both unsuspected, and unattended by any significant tissue damage. Crystal deposition in joints can also be clinically silent; many cases of 'chondrocalcinosis' (calcification of articular cartilage) are asymptomatic, without there being any clinical or radiological evidence of joint damage, although in other instances, the deposits apparently lead to accelerated, severe joint destruction. Even the gout crystal (monosodium urate monohydrate) has been found in asymptomatic joints with no evidence of inflammation.

Thus as well as understanding crystal deposition it is important to investigate the processes leading to tissue damage. These are usually complex, and interrelated with other pathological processes, so that it is difficult to identify a single event precipitating a specific disease. For example, many different reactions are activated by urate crystals during an acute attack of gout; some of the resulting changes, such as the lowered tissue pH, may result in further crystallization, whereas others, such as increased temperature may aid crystal dissolution. Furthermore, urate and other crystals apparently activate anti-inflammatory as well as pro-inflammatory pathways, so that acute attacks are self-limiting.

It is useful to divide crystal effects into the two general categories shown in Fig. 5.1. (1) The *mechanical* effects produced by crystal deposits are often obvious; they include the tube blockage and local tissue destruction caused by large crystal masses such as gall-stones or kidney stones; and the loss of elasticity and strength of flexible tissues such as arterial walls or joint cartilage. (2) The second category includes the *biophysical and biomechanical* effects associated with the presence of individual crystals. These include the direct effect of crystal surfaces acting as foreign chemical entities in the body. They can bind and denature proteins, thus activating phagocytosis, inflammation and other pathological processes. Crystal surfaces of the right type, shape and size may also interact with, or disrupt, cell membranes, alter cellular metabolism and lead to the release of proteolytic enzymes or

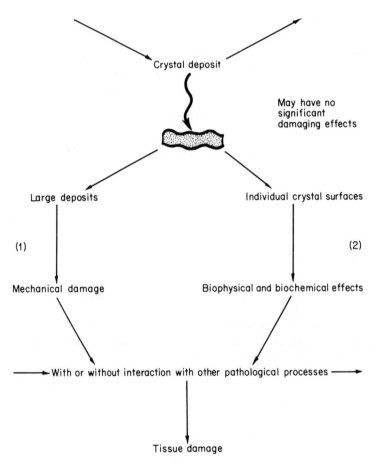

Fig. 5.1 Tissue-damaging effects of crystal deposits.

inflammatory mediators. Of the many types of crystal-induced damage, the acute inflammatory response is particularly obvious and well investigated. In recent years our understanding of this process has expanded enormously, and as a result we can look forward to new understanding of old therapies (such as the very effective and remarkably selective action of colchicine on gouty inflammation) as well as to new therapeutic avenues. Research on the mechanism of chronic crystal-induced tissue damage is only just beginning, and may even provide ways of preventing joint destruction in some types of osteoarthritis.

5.2 The crystal and its environment

The fact that similar crystal deposits may have different effects in different tissues and in different people, points to the importance of considering both the crystal and its surroundings when studying a disease. This leads to a 'seed and soil' concept (e.g. the parable of the sower, *Luke* **8,** 5); where the tissue (the 'soil') is important in determining both the development of the deposit (the 'seed') and in its effects. In this section we will discuss the important tissue and crystal factors independently, before going on to discuss how they interact.

5.2.1 The crystals

The mechanical and biochemical properties of crystals can be considered separately (Fig. 5.1). Mechanical effects arise from the presence of large, hard structures, whereas the biochemical reactions depend on the available surface, and are therefore more influenced by the physical and chemical properties of the crystals.

(1) As mentioned above, large crystal deposits can give rise to disease through simple *mechanical effects* such as blocking ducts or by hardening and weakening flexible tissues.

In the case of *stone formation*, any hard, relatively insoluble lump is as effective as any other in blocking a duct, and it is the size and position that determine the resulting disease. However, the means by which large stones go on increasing in size in collecting ducts (gall stones or kidney stones), so as to cause eventual obstruction, is not well understood. They are aggregates of crystals, often of several different chemical species, and formation presumably continues through either epitaxial nucleation and growth on crystals already present, or by separate, new nucleation sites forming in the organic matrix or surrounding fluid and growing to become entangled with and engulfed in the earlier deposits (see Chapter 3). Whatever the mode of growth, the interaction between the tissue matrix and the growing crystalline aggregates must be important. The extent to which crystal surfaces bind polysaccharides is probably important in determining whether aggregation occurs, or whether the small crystallites are simply washed out of the system. Further understanding of this accumulation of stones, and subsequent tube blockage, may help us to develop further medical means of dispersing stones or preventing the formation of obstructions. Duct blockage leads to back-pressure and tissue damage; it also predisposes to infection, and infection may further alter the tissue matrix predisposing to further crystal deposition particularly by changing the local pH. Thus infection is a well-known aetiological agent in renal stones, reemphasizing the interaction between stone formation and tissue damage.

Crystal deposits also alter the *mechanical properties* of tissues in which they are formed. Almost all of the chemicals which form into crystals in the body are either ionic, like sodium urate, or strongly hydrogen-bonded like cholesterol. In either case, the binding forces within the crystal are strong, and as a result the deposits are considerably harder than the soft, elastic tissues surrounding them. Hence loss of elasticity and compliance will occur in structures such as arterial walls or articular cartilage once they become calcified. The effect will occur in proportion to the volume and distribution of the crystal deposit; if the crystals are uniformly distributed hardening will be the main effect, however, if they tend to occur as localized loose aggregates they may also produce significant weakening of the structure, resulting in local fracture and possible release of the crystals.

(2) The *biochemical* effects of crystals are much more complex, and several factors determine the nature of the inflammation, fibrosis and other tissue reactions to crystals (Table 5.1).

Table 5.1   Features of crystal deposits which determine crystal damage

---

(a) Mechanical effects of macroscopic deposits depend on:
Stone formation
(1) Blocking ducts, leading to back-pressure, and infection.
(2) Destroying the tissue matrix.

Hard crystals in soft tissues
(1) Hardening tissues, and altering elasticity.
(2) Causing local fractures and weakening.

Wear, surface abrasion by hard particles.

(b) Biochemical effects of individual crystals depend on:
Crystal size
Crystal shape
Crystal surface chemistry
(1) Charged ions interacting with other chemical groups.
(2) Surface 'roughness' and energy.

The size of individual crystals is known to be important in acute inflammation. For example, different sized urate crystals cause a varying degree of oedema after being injected into a rat's paw (Table 5.2). 'Amorphous' deposits (i.e. very small particles), and large crystals (> 20 μm long), are relatively ineffective, whereas crystals of about 5 μm in length (similar in size to those found in the synovial fluid in gout) are most reactive. Shape may also be important, acicular crystals may be the most phlogistic; grinding crystals reduces their inflammatory potential as well as altering their shape although there may be other reasons for this, and it is difficult to assess the effect of shape alone. In experimental work it has been difficult to separate the effects of size and shape from alterations in the total surface area available, or in the nature of the crystal's surface.

Table 5.2 Increase in diameter of rat paw 24 h after injection of 5 mg urate crystals in 0.2 ml sterile saline.

	Mean $\pm$ S.E., six recordings (mm)
'Amorphous' urate (particles <0.1 μm long)	0.10 ± 0.03
'Small' crystals (about 5 μm long)	0.38 ± 0.04
'Large' crystals (> 20 μm long)	0.26 ± 0.03
Ground small crystals (shape and surface altered)	0.23 ± 0.05

The chemical nature of the crystal surface is certainly the most important factor determining its biochemical effects. Specific examples are given below, but it is worthwhile first discussing the general nature of surface interactions.

Fluid-phase tissue areas, such as the synovial joint space, contain water, a mixture of small ions, macromolecules such as polysaccharides and proteins, and cells. The first surprising thing about this structure is that it does not precipitate out polymers and crystals more extensively. A random synthetic mixture of similar components would probably form a precipitated mass below a watery liquid, and the fact that this does not happen implies that surface-active factors are present, allowing macromolecules and surfaces to co-exist in a non-precipitating system. This is bound to be important when considering the introduction of a new surface such as a crystal.

Proteins can be thought of as a hydrophobic (fatty) core surrounded by an ionized exterior layer which interacts with water (hydrophilic). A mixture of different proteins will precipitate unless the molecules repel by being charged with the same polarity as the cells and polysaccharides and as each other. Globulins, cells and many other macromolecules do indeed possess a net negative charge and so will be mutually repulsive. The introduction of a charged crystal, especially if it is positively charged might be expected to result in strong interactions and aggregation although it is interesting to note that most of the crystals and other particles involved in disease are believed to possess a net negative charge similar to that of the cells and proteins.

In addition to charge interactions, the various molecules in the system also interact by hydrogen bonding, dipolar attractions and dispersion forces (Fig. 5.2). The net result of adding a crystal surface is a summation of all these various interactions; and it is too complicated to predict what will

1. Charge Interaction

Crystal with net negative charge

Macromolecule with net positive charge

2. Hydrogen bonding

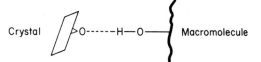

3. Dipolar attractions

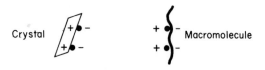

4. Dispersion forces (Van der Waals forces)

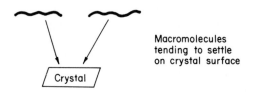

Macromolecules tending to settle on crystal surface

Fig. 5.2 Possible surface interactions between crystals, macromolecules and cell membranes.

happen in, for example, synovial fluid. Indeed the problem is almost insoluble even for a crystal in a pure solution of its own ions. However, the species present can be divided into three main groups: polar hydrophilic molecules capable of hydrogen bonding; non-polar, hydrocarbon, hydrophobic molecules; and amphibiles (such as detergents) where part of the molecule is hydrophilic and part hydrophobic. Most globular proteins, as well as cell membranes, are amphibiles, with the outermost surface containing the hydrophilic groups that interact with the surrounding water. Crystals may have exposed surfaces which are relatively hydrophobic. These should adsorb the hydrophilic parts of proteins and membrane components, causing these structures to denature.

Another good example of the effect of surfaces in biology is seen in the search for non-thrombogenic materials for surgical implants. The preferred materials are those with either non-polar, hydrophobic surfaces (e.g. flurocarbon polymers) or with highly polar surfaces (e.g. pyrolytic carbon). Platelet adhesion seems to occur on any surface which absorbs gammaglobulin and fibrinogen in preference to albumin, but there is no good way of deciding which factors predetermine this preference. An understanding of the surface interactions has a wide-ranging relevance to many areas of medicine, in addition to the crystal deposition diseases.

Not all crystals cause inflammation (Table 5.3). Diamond crystals have little or no effect in any of the standard *in vivo* and *in vitro* tests of inflammation and its mediation. Even more interesting is the fact that some crystals which are chemically very similar have quite different phlogistic potential; L-cysteine crystals have no effect, and stishovite and anatase are also non-inflammatory, whereas other silicon dioxide crystals, such as quartz and cristobalite cause a brisk reaction. This suggests that inflammation is very dependent on the surface configuration of the atoms.

Similarly, non-crystalline particles are sometimes inflammatory and sometimes inert. Latex beads and carbon particles are phagocytosed by

Table 5.3   Inflammatory and non-inflammatory crystals

Inflammatory	Non-inflammatory
Monosodium urate monohydrate	L-cysteine
Calcium pyrophosphate dihydrate	
Dicalcium phosphate dihydrate	Diamond
Hydroxyapatite	
Cholesterol	
α-Quartz (SiO_2)	Stishovite (SiO_2)
Tridymite (SiO_2)	Anastase (TiO_2)
Cristobalite (SiO_2)	

cells, but cause little or no release of inflammatory mediators, and cause no inflammatory reaction *in vivo*. Conversely, particles from plant thorns are very inflammatory. We should probably be investigating a wider range of particles than the crystals involved in joint disease if this sort of inflammation is to be understood.

Understanding is also hampered by the difficulty in measuring the surface area of a particle, and knowing the surface atomic configuration and what chemicals are bound to it. Our experiments relating the electrophoretic mobility of particles (i.e. their net charge property in physiological saline) to inflammatory mechanisms are among a small number of rather crude methods of looking at what factors make some particles inflammatory, and others inert.

5.2.2. *The environment*

The surroundings of a crystalline precipitate can effect the subsequent development of disease in a number of ways. First, it will determine the morphology, size and distribution of the deposit by means of local variations of salt concentration, temperature, nucleating surfaces, growth poisons and disruptive mechanical factors, as outlined in Chapter 3. As previously discussed, these local tissue factors must decide the peculiar distribution of crystalline deposits in conditions like gout and pyrophosphate arthropathy, and thus the nature of the disease.

Other ways in which the site of the deposit effects tissue damage are often self-evident. For example, a tophus of urate crystals in a joint causes considerable damage, whereas a deposit of the same size in the ear-lobe may go unnoticed for years. Similarly, a stone in the gall bladder may be quite innocuous until it shifts position and blocks the common bile duct.

However, the core of the 'seed and soil' concept is that particle-induced reactions such as acute crystal-induced inflammation are very dependent on the state of the tissues surrounding the deposit. It would appear that inflammation is unlikely to occur if the crystals are embedded in a firm tissue matrix like cartilage, where they are not exposed to circulating tissue fluids or phagocytic cells, and cannot activate any inflammatory system. Acute inflammation will depend on crystals coming into contact with one of the protein or cellular systems that can trigger the response. Small, subtle changes in the environment of a crystal deposit can be enough to activate a damaging process, resulting in a sudden attack of pain, and the patient may have no explanation for the timing of its onset.

Acute calcific periarthritis is a good example of this (Chapter 8). Deposits of crystals may be present in the periarticular soft tissues for years without causing trouble; a mechanical or metabolic change can disrupt the deposit, releasing microcrystals of hydroxyapatite. If they activate inflammatory proteins, or undergo phagocytosis, and if their numbers and rate of release are appropriate, acute inflammation follows. Thus a patient may have many deposits, and be asymptomatic; some of these deposits may get disrupted, but not cause problems, while on another occasion an intense inflammatory response may occur, and the patient presents to his physician with sudden severe pain without an obvious cause.

It can be seen that both the crystal and its environment affect the pathological process. In the next section, the specific pathways implicated in a variety of individual diseases will be considered.

5.3 Mechanisms of crystal-induced joint disease

Chapters 6–9 describe several diseases of the joints associated with the presence of crystals. The main features of these conditions are inflammation and cartilage destruction. A number of mechanisms thought to be instrumental in these pathological processes are outlined (Table 5.4).

Table 5.4 Probable mechanisms involved in the production of crystal-induced joint damage

---

(1) Acute inflammation
  (a) Activation of protein/enzyme systems (e.g. Hagemann factor and complement)
  (b) Interaction of crystals and cell membranes (platelets, leucocytes)
  (c) Crystal phagocytosis
(2) Chronic inflammation
  Persistence of stimulus to acute inflammation plus activation of fibrosis and other chronic inflammatory mediators.
(3) Cartilage and bone destruction
  (a) Physico-chemical alterations in the articular cartilage
  (b) Products of inflammatory reactions
  (c) Surface abrasion

Of the various mechanisms listed, only the acute inflammatory response has been investigated in detail and most of the experiments have been done with urate crystals. More work is needed on the other mechanisms of crystal-induced damage, and on particles other than urate crystals.

5.4 The acute inflammatory response

Acute inflammation is the most obvious and dramatic consequence of the presence of crystals in joints. Gout, pseudo-gout and acute calcific periarthritis are all good examples. At first sight, these three conditons have a number a similarities, all being dominated by a severe, self-limiting, crystal-induced inflammatory response. There are, however, differences in the time course and therapeutic responses of these diseases, as well as in the crystals found in the joints. Research using *in vivo* models of inflammation, and *in vitro* tests have both shown that several separate mechanisms may be involved in inflammation and that different crystals will activate different pathways; it is perhaps not surprising therefore to find that gout and pseudo-gout respond differently to drugs or that a gouty attack can be quite different in separate patients, or on different days.

Before examining the mechanisms involved, we will discuss correlations between crystal structure and the intensity of the inflammatory response as a whole, because one of the most interesting facts to emerge from animal testing is that crystals that are superficially very similar, have quite different inflammatory potential. A list of inflammatory and non-inflammatory minerals was given in Table 5.3. On a weight-for-weight basis, our experiments suggest that urate causes more inflammation than pyro-phosphate, hydroxyapatite or brushite crystals, although it has not been possible to compare the same number and size of crystals. However, other, similar crystals cause virtually no reaction at all when applied to animal models such as rat-paw oedema. Thus L-cysteine is non-inflammatory, and experiments on modified silicon and titanium dioxides show that some are inflammatory, and some are not. *In vivo* experiments have also shown that the response is crystal specific, and that many very fine or 'amorphous' materials are not inflammatory. To what extent this depends on size, surface or shape of particle remains difficult to ascertain.

The structural basis of this differing inflammatory potential has been investigated by Mandel (1976). He has concluded that the key factor is not the presence of hydrogen-bonding groups, as has been suggested by others, but the surface roughness on an atomic scale, thus 'smooth' crystals like cysteine are inactive, whereas urate, with many active charged groups sticking out from the surface is very phlogistic. In so far as this is related to surface energy, which will control absorption phenomena, this seems most reasonable. A corollary of this is that chemical treatments of the crystal surface may render it less active, and this is being explored in a number of laboratories.

The electrophoretic mobility of crystals has been studied by several people. If particles are suspended in salt solutions, and a current applied, any surface charge will result in movement towards one or other pole. This movement can be measured by a variety of techniques, including light microscopy, or the Doppler shift in frequency of a laser beam reflected by the particles. By varying the salt concentration, the net surface charge density, or 'zeta potential' of the particle can be estimated. Inflammatory particles have a net negative zeta potential, similar to that of erythrocytes. This charge, which presumably reflects the surface activity of the crystals, correlates with their ability to induce acute inflammation in animal models; in the case of both crystals and plant thorns, a higher zeta potential is associated with the generation of more inflammation. The mechanism of this effect is discussed further in the next section; and it shows the importance of the surface of particles in the production of joint damage.

Various pathways involved in mediation of the acute inflammatory attack are shown in Fig. 5.3. Release of free crystals into the tissue space is probably the first event, and what follows will depend on the population of proteins,

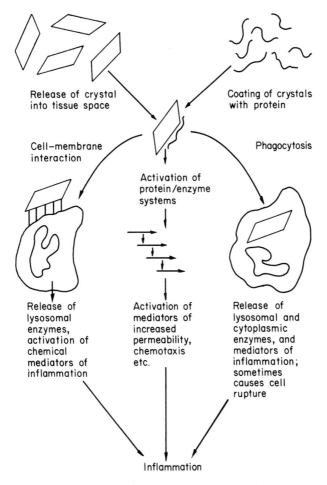

Fig. 5.3 Possible steps involved in crystal-induced inflammation.

which coat the crystal and may be activated by them and on the cells present, which may also interact with, or phagocytose, the particles.

5.4.1   Protein binding

The first important step following addition of a crystal to a medium like synovial fluid must be *protein binding*. *In vitro* experiments show that urate and pyrophosphate crystals are avid binders of protein, especially im-munoglobulin, IgG (Kozin and McCarty, 1976). In view of the net negative charge of the crystals, the preference for IgG over IgM coating may be due to the lower negative charge of the IgG. However, the effect is not specific, and crystals also have an affinity for many other proteins; it has even been suggested that absorption of bacterial endotoxin may be important in the inflammatory response in some cases. There is also *in vivo* evidence for protein coating; immunofluorescent staining of gouty tophi has shown evidence for the presence of IgG and IgM on the crystal surface and scanning electron micrographs of urate or pyrophosphate crystals recovered from synovial fluid usually show the presence of a granular coating on the surface (see Fig. 6.11).

Protein binding is likely to change the surface characteristics of the crystals, and, as explained below, may enhance interactions with cell membranes, phagocytosis, and activation of cell-free inflammatory mechanisms.

Cell-free protein systems that can be activated by crystals include the kinin and clotting systems (via Hagemann factor) and the complement cascade. Most of the work on these systems has been with urate crystals, and what follows refers to this; to what extent other particles have the same effect remains largely unknown.

5.4.2   Complement activation

Complement activation has been studied by several groups. The comple-ment system comprises many different serum proteins, normally present in an inactivated state; activation occurs by cleavage, and by uncovering enzymatic sites on these proteins, resulting in a cascade of events producing factors that are chemotactic, enhance phagocytosis, and cause lysis of cell membranes (Fig. 5.4). Activation by immune complexes has a central role in the immunological response, but Hasselbacher (1979) and others have shown that urate crystals can also activate the classical complement pathway, even in the absence of IgG binding, although surface proteins may enhance the effect. As explained below, complement levels are reduced during gouty inflammation. Also prior complement depletion reduces the degree of inflammation subsequently induced by urate crystals, suggesting that complement activation is important in gout.

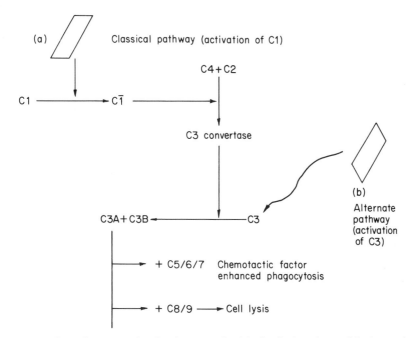

(a) Classical pathway (activation of C1)

C4 + C2

C1 ⟶ C̄1 ⟶

C3 convertase

C3A + C3B ⟵ C3

(b)
Alternate
pathway
(activation
of C3)

+ C5/6/7 Chemotactic factor
enhanced phagocytosis

+ C8/9 ⟶ Cell lysis

Fig. 5.4 Complement activation by crystals: (a) classical pathway (b) alternative pathway.

Recent experiments in our laboratory have shown that inflammatory crystals can also activate the alternate pathway of complement (Doherty *et al.*, 1982). This occurs in immunodeficent, as well as in normal serum, and may not depend on protein coating. Complement activation can be induced by plant thorns and other phlogistic particles, but not by diamond crystals and other substances that will not induce inflammation *in vivo*. Our results have shown some correlation between complement activation, surface charge of the particles, and foot-pad swelling in rodents; this suggests that complement may be an important mediator of inflammation, and that the net surface charge of a particle may relate to its ability to cleave complement C_3.

5.4.3 *Hagemann factor*

Hagemann factor (factor XII of the clotting system) is another protein that can be activated by crystals. Recent research has shown that it is split into two components by urate crystals; one remains bound to the crystals, and the other is a free serum factor. Once activated in this way, Hagemann factor can initiate at least two series of chemical reactions; one is the clotting system, the other is the kinin system, leading to formation of potent vasodilators. This pathway could enhance the inflammatory response, although its role *in vivo* is not clear, and chickens, which lack Hagemann factor, produce a brisk inflammatory response when challenged with urate crystals.

5.4.4   Cell membranes

The interaction between crystals and cell membranes has been studied by many authors. Experiments using labelled red cells, incubated with crystals, have shown that membranolysis occurs. That this depends on the crystal surface is shown by the lack of effect of some minerals, and by the fact that alteration of the surface by grinding or other manipulations decreases membranolysis. Of the crystals primarily involved in acute joint inflammation, urate has the largest effect, followed by pyrophosphate, hydroxyapatite and brushite crystals, in that order. Interactions between crystals and cell membranes can also alter cellular metabolism, and cause secretion of inflammatory mediators; thus urate crystals induce release of serotonin and other mediators from platelets, and lysosomal enzymes are discharged from leucocytes in contact with crystals in the absence of phagocytosis. It is interesting to speculate whether other cells may also be metabolically modified when in contact with crystals.

The mechanism behind crystal–membrane interactions has been investigated by Weissman and his colleagues (1971) and Wallingford and McCarty (1971). Using liposomes as model cells, and erythrocytes, they have shown that binding and membranolysis are inhibited by hydrogen acceptors such as polyvinyl pyridine-N-oxide. This, and other evidence, led to the hypothesis that hydrogen donor sites on the crystals attach to acceptor sites on the outer surface of the lipid bilayer of cell membranes. However, much of this work is based on non-physiological *in vitro* experiments, and the influence of protein binding and crystal surface charge on membrane binding and rupture deserves more study. Many authors believe that the attachment of particles to cells, and activation of membrane enzymes, must depend on membrane receptors being activated by protein attached to the cell or the particle. The IgG Fc receptor may well be important (see below). Interestingly, addition of serum inhibits erythrocyte membranolysis by crystals, re-emphasizing the importance of the protein milieu of the site.

5.4.5   Phagocytic cells

Phagocytic cells have a central role to play in crystal-induced inflammation. There are two types of evidence for this statement: first depletion of phagocytic cells reduces the inflammatory response in experimental animals; and secondly, morphological studies show a predominance of polymorphonuclear cells, with active crystal phagocytosis, in crystal-induced inflammation (Phelps and McCarty, 1966).

Several authors have documented the events that occur on addition of urate crystals to phagocytic cells *in vitro*. There is morphological evidence of active phagocytosis within a few minutes. Once inside the cell the crystal is surrounded by a membrane (phagosome), lysosomes may fuse with the membrane (Fig. 5.5), although the membrane subsequently disappears and cell degranulation occurs. Finally the cell may rupture and the crystal is released into the culture medium to be re-phagocytosed. A similar train of events has been observed with pyrophosphate and hydroxyapatite crystals,

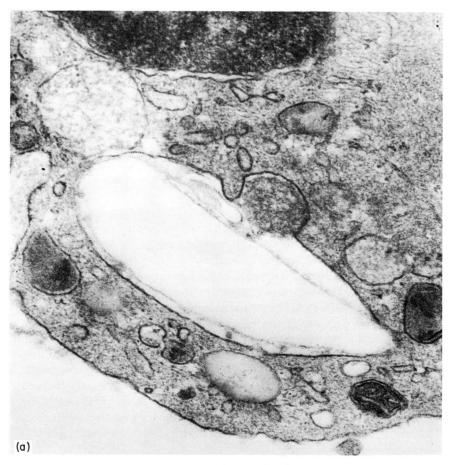

(a)

Fig. 5.5 (a) Transmission electron micrograph of a phagocytic cell that has internalized a urate crystal. Note the fusion of a lysosome with the membrane surrounding the crystal cleft to form a phagolysosome (× 30 000).

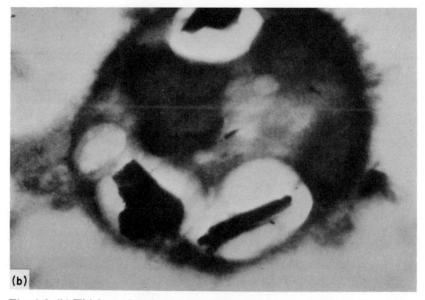

Fig. 5.5 (b) Thick-section electron micrograph showing crystals within a phago-cytic cell (× 12 000).

although the process of cell degeneration and cell rupture take longer, and far fewer cells die.

The reason for this observed train of events is not yet clear. On the basis of experiments quoted above that show preferential binding of IgG to crystal surfaces, and inhibition of red cell membranolysis by hydrogen group acceptor substances, the sequence illustrated in Fig. 5.6 has been proposed. It is suggested that the crystal binds to the Fc receptor on phagocytic cells via its IgG coat. This activates phagocytosis, and phagolysosomes form as for ingested bacteria. However, if the protein coating is then stripped from the surface by enzymic action, membrane–crystal binding by hydrogen bonding could occur leading to fracture of the phagolysosome membrane, and then the cell membrane itself.

The problems with this argument include the fact that phagocytosis can apparently take place in the absence of IgG, and that positively charged lysosomal enzymes might be expected to remain attached to the negatively charged crystal, thereby avoiding any direct contact with the membrane. However, the morphological events are undisputed (Schumacher, 1977; Rajan, 1975).

In vivo, phagocytic cells lining the synovial membrane may also make an important contribution to these diseases, and there is evidence that they do phagocytose crystals from the joint space.

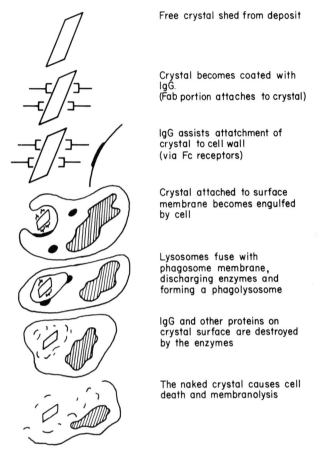

Free crystal shed from deposit

Crystal becomes coated with
IgG.
(Fab portion attaches to crystal)

IgG assists attatchment of
crystal to cell wall
(via Fc receptors)

Crystal attached to surface
membrane becomes engulfed
by cell

Lysosomes fuse with
phagosome membrane,
discharging enzymes and
forming a phagolysosome

IgG and other proteins on
crystal surface are destroyed
by the enzymes

The naked crystal causes cell
death and membranolysis

Fig. 5.6 Proposed mechanisms of crystal phagocytosis.

As in complement activation, the importance of the net charge property
and other aspects of the surface of crystals is being investigated further, and
may lead to more understanding of protein binding and cell membrane
interactions. These experiments may also suggest alternative lines of
therapy such as coating crystal surfaces, as mentioned in Chapter 11.

5.4.6 Mediators

The final common pathways of all inflammatory responses include
vasodilation and cellular infiltration. The *mediators* of these events in
crystal-induced lesions have been studied. There is evidence for an
important role for histamine, kinins and prostaglandins, in that order of
release, in these as in other inflammatory lesions. There is also good

evidence that complement components are activated to aid vasodilations and chemotaxis. Cells that have phagocytosed crystals also release a chemotactic factor into the surrounding medium (Spilberg *et al.*, 1973), and both lysozomal and cytoplasmic enzymes.

Examination of synovial fluids in acute crystal-induced reactions such as gout, shows the presence of these cell-derived chemical mediators. Exudates from experimental crystal-induced inflammatory reactions, and from preparations of cells and crystals cultured together can also be analysed. Prostaglandins and enzymes such as lactate dehydrogenase (from the cytoplasm) and glucuronidase (from the lysosomes) are released (Fig. 5.7). A specific chemotactic factor has also been described, a small molecular-weight protein released from leucocytes incubated with crystals, with very potent chemotactic activity. The role of these various mediators is not apparent, but as in all inflammatory reactions, several different reactions probably coexist and interact in response to crystals.

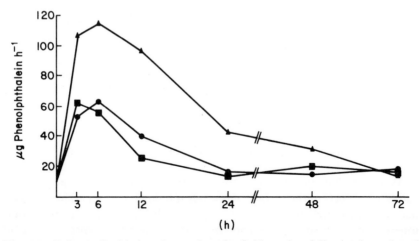

Fig. 5.7   Release of acid phosphatase into the fluid exudate following intrapleural injection of crystals in the rat. Urate, pyrophosphate and apatite crystals all cause acid phosphatase release from cells, maximal 6 h after crystal injection (after Glatt *et al.*, 1979).

So far, this chapter has described some of the individual mechanisms contributing to acute inflammation. The importance of the different pathways almost certainly varies in different situations and species. Animal models such as the avian microcrystal arthritis test, dog knee synovitis, and rodent food-pad swelling and pleurisy may help to sort out which factors dominate the reaction. They suggest that crystal phagocytosis with subsequent release of enzymes and inflammatory mediators, plus direct complement activation are most important. In synovial joints a special factor enters into the equation, the presence of phagocytic lining cells designed to remove debris from the joint. They may ingest the crystals, and be a major factor initiating the synovitis. Other aspects relevant to particular diseases such as acute gout will also be mentioned in the next chapters.

5.5 What starts and stops acute inflammation?

A characteristic of all crystal-induced inflammatory reactions is their sudden onset, quickly becoming maximal, and their self-limiting nature. Hippocrates knew that 'in gouty affections inflammation subsides within 40 days' (*aphorism* **VI,** 40), and it usually takes much less time. So why do these reactions suddenly develop, and just as quickly subside?

In the next three chapters possible mechanisms specific to gout, pseudogout and calcific periarthritis are mentioned. Table 5.5 outlines the basic possibilities.

Table 5.5   What starts and stops acute inflammation?

(1) Possible causes of onset
    Crystal shedding from the deposit
       (trauma or dissolution of deposit)
    Acute crystallization
(2) Possible causes of limitation
    Removal of crystals
    Activation of anti-inflammatory pathways

Acute inflammation could be triggered by release of crystals from a pre-formed deposit, or by acute crystallization. The reaction presumably depends on the delivery of a sufficient number of crystals of the right surface, size and shape, to the site of the inflammation. Most authors have favoured crystal shedding as the most likely way of fulfilling these criteria. Particles might be released from a deposit by mechanical disruption, or by partial dissolution. Many of the clinical situations which induce gout and pseudogout involve trauma or a metabolic disturbance that might be expected to solubilize the crystals.

An attack might be terminated by removal of crystals, or by activation of inhibitory mechanisms. Although little is known about either of these possibilities, an important clinical observation is that crystals can still be found in the synovial fluid after an attack of gout or pseudogout has terminated. The number or nature of the crystals may have been altered sufficiently to render them inactive, but it seems just as likely that some, as yet unknown, anti-inflammatory mechanisms are being activated to stop the attacks.

5.6   Chronic inflammation and fibrosis

5.6.1   Chronic inflammation

There is no strict dividing line between acute and chronic inflammation. Many crystal-induced diseases such as gout, pseudogout and calcific periarthritis, are characterized by self-limiting inflammation. The attacks stop, even in the continued presence of the crystalline stimulus. However, there is both clinical and experimental evidence for crystal-induced chronic inflammatory responses as well. Acute inflammation is characterized by vasodilation and influx of polymorphonuclear cells. By contrast, chronic inflammatory lesions contain macrophages, and evidence of fibrosis and new vessel formation. These chronic pathological changes have been observed around synovial and soft-tissue deposits of crystals, although they are not always present. Giant cell formation is sometimes seen as well.

Animal experiments have shown that intradermal injection of crystals in rodents can cause a chronic 'granulomatous' reaction lasting for several weeks.

Hydroxyapatite and pyrophosphate crystals proved more potent in this respect than urate. This is in contrast to the acute inflammatory response, where weight-for-weight urate is the most active crystal. Crystals can also be injected intra-dermally in man. This results in an area of erythema and induration that is maximal about 24 h after the injection (Figs 5.8 and 5.9). As this initial inflammation subsides, a small granuloma forms; a biopsy (from the author's arm) showed a chronic granulomatous reaction to the crystals, with some giant cells and fibrosis around the edge of the lesion. The time course of this reaction in human skin is not dissimilar to that of an immunological reaction such as a Mantoux test; it has been used to assess anti-inflammatory drugs, and to investigate inflammatory reactions to crystals in different patients.

5.6.2   Fibrosis

Fibrosis in response to crystals has been investigated by Allison and his colleagues (1978) and others. Increased production of collagen by the fibroblasts is a part of all chronic inflammatory responses, and can occur in the absence of a great deal of inflammation. This is particularly important in crystal-induced lung fibrosis as in silicosis and asbestosis.

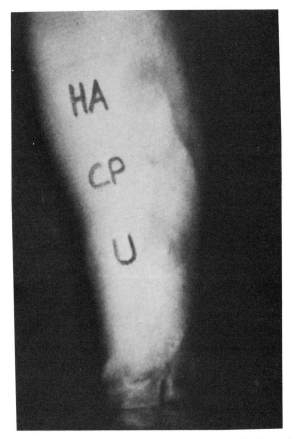

Fig. 5.8 Inflammatory reactions in human forearm skin 24 h after injection of 5 mg of hydroxyapatite (HA), calcium pyrophosphate dihydrate (CP) and urate (U) crystals in sterile saline. Chronic granulomas form if more crystals are injected.

Inhaled particles of less than 5 μm diameter can reach the terminal airways and are then ingested by alveolar macrophages, There are three possible consequences of this, depending on the nature of the particle: first no reaction may occur (as with carbon particles); secondly, a chronic inflammatory response is set up (e.g. asbestosis); and thirdly, fibrosis can occur (e.g. silicosis). The importance of the crystal surface is again apparent, and removal of surface magnesium groups from asbestos fibres blocks their ability to mount a reaction. Silica and asbestos particles are both cytotoxic as well as fibrogenic, and two mechanisms, also dependent on surface structure, have been identified. In the case of silica, hydroxyl groups on the surface are thought to be hydrogen-bonded to membrane lipids, whereas the magnesium surface groups of asbestos probably form electrostatic

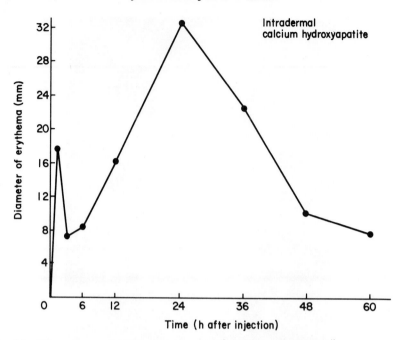

Fig. 5.9 Time course of the erythema occurring in human forearm skin in response to injection of 10 mg hydroxyapatite crystals in 0.2 ml sterile saline. Saline alone or amorphous calcium phosphates cause the initial reaction (a few minutes after injection) only. Crystals, including hydroxyapatite, urate and pyrophosphates cause a secondary peak of inflammation 24 h after the injection.

bonds, resulting in an ion-conducting channel between the crystal surface and the cell membrane.

The mechanism behind silica- or asbestos-induced lung fibrosis has been explored further. *In vitro* experiments show that macrophage ingestion results in the release of a fibrogenic factor that then stimulates fibroblasts to increase collagen production. The response is dependent on the dose and size of the particles, and once more demonstrates the remarkable ability of cells to respond differently in various sites, and in response to different particles. Fibrosis sometimes occurs around crystal aggregates in or around joints, and may prevent crystal release and access of cells or protein systems that activate the acute inflammatory response. A granulomatous reaction, with fibrosis, also occurs around many tophi.

5.7   Destruction of articular cartilage and bone

Acute inflammation is one of the two chief reactions associated with the presence of crystals in joints. The other is a chronic destructive arthropathy with damage to the articular cartilage and bone.

Some of the possible mechanisms involved are shown in Table 5.4. The simplest hypothesis suggests that joint damage results from the products of repeated acute inflammatory events or low-grade chronic synovitis. Proteolytic enzymes are certainly produced by the synovium, and deleterious inter-cellular communicators such as 'catabolin' may also be important. However many patients have repeated acute inflammatory events and little chronic damage, e.g. the sub-group of young men with chondrocalcinosis and pseudogout (see Chapter 7), and it seems likely that other mechanisms are also involved in cartilage destruction. Some recent experimental work has suggested one possible mechanism of cartilage destruction. Crystals incubated with cultured cells from synovial membrane have been found to cause release of proteases including collagenase, and PGE_2. The proteases may contribute to damaging the cartilage matrix, and factors such as PGE_2 may activate bone resorption. A few patients in whom synovial fluid hydroxyapatite crystals and high levels of activated collagenase coexist have been described by McCarty (1981); the joints were severely damaged, suggesting that this mechanism may operate *in vivo*.

In many cases of chondrocalcinosis a severe destructive arthritis develops relatively quickly, without much overt inflammation. This may also be due to a direct damaging effect of the presence of crystal deposits in the cartilage matrix. As already mentioned these deposits must alter the compliance of the structure. In response to load, this could have one of two consequences: (1) it might reduce the load-spreading ability of the cartilage, transmitting excess forces locally to the bone below; or (2) fractures could develop in the cartilage at the boundary between the relatively hard, inflexible deposit and the surrounding matrix. Surprisingly little experimental work has been done on this subject.

An idea of the forces applied to some intra-articular crystal deposits can be seen from electron micrographs of calcified menisci, showing apparent fracturing of the individual crystals, as well as of the cartilage (Fig. 5.10).

Another possible mechanism of crystal-induced cartilage destruction is surface wear. Hard, sharp, crystalline particles could provide an ideal cutting surface to abrade the surface of hyaline cartilage. The size of the particles might be important, as in sand-paper (coarse paper is more abrasive than fine grade papers). Again there is little direct evidence for this mechanism, but two research techniques mentioned elsewhere in this book could be relevant. Ferrography has shown the range and extent of the particles found in synovial fluid, and analytical electron microscopy has revealed the presence of some calcium phosphate crystals embedded in the surface of articular cartilage.

Bone destruction could result from loss of articular cartilage and the

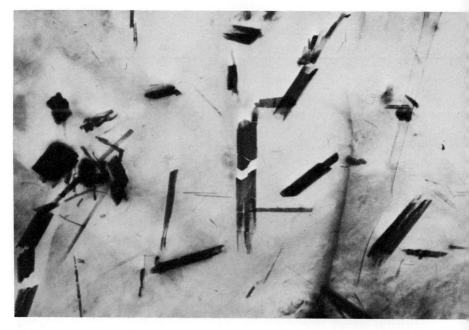

Fig. 5.10   Calcium pyrophosphate dihydrate crystals in articular cartilage viewed in the transmission electron microscope. Fractures of the crystals are clearly visible, and could be due to a processing artefact, or to the mechanical forces applied to the area *in vivo* (× 6000).

normal load-bearing and stabilizing features of the synovial joints. The erosions of gout may be in part due to stress concentrations at the bone surface caused by hard zones in the cartilage but the possibility exists that crystals can also activate some cellular mechanisms altering subchondral bone.

The mechanisms behind the chronic destructive joint diseases associated with crystals deserve more study. These conditions represent a much more relevant and difficult clinical problem than the self-limited inflammatory reactions.

5.8   Summary

Inflammatory responses to crystals are well documented and many of the pathways involved are apparent. A complex series of events can take place, depending on the exact nature of both the crystal, and the surrounding proteins and cells.

Other mechanisms of crystal-induced damage are less researched, although fibrosis, tube blockage and altered tissue compliance, can all result in overt disease expression.

Most attention has been focused on the joints, although crystal deposition is a feature of many diseases in other parts of the body. The mechanisms involved in other organs are probably similar to those occurring in joints, although the ability of crystals to bind proteins, and affect cell membranes, give them a vast potential for inducing subtle or overt changes in immunology and cell metabolism throughout the body. As our understanding of pathology increases, the role of microcrystals in other major diseases, such as atheroma, may assume increasing importance.

Further reading

DIEPPE, P.A. and DOHERTY, M. (1982) Particle induced disease, in *Recent Advances in Pathology: Bone and Joint Disease*, Springer Verlag, Berlin.

DUNN, C.J., DOYLE, D.V. and WILLOUGHBY, D.A. (1978) Experimental methods in the study of Crystal Deposition Diseases *Eur. J. Rheum. Inflam.* **1** (2), 135.

MALAWISTA, S. (1977) Gouty Inflammation. *Arth. Rheum.* **20,** 241.

McCARTY, D.J. (1970) Crystal-induced inflammation of the joints. *Ann. Rev. Med.* **21,** 357.

SCHUMACHER, H.R. (1977) Pathogenesis of crystal-induced synovitis. *Clin. Rheum. Dis.* **3** (1), 105.

Text references

ALLISON, A.C. (1978) Inflammatory and fibrogenic effects of tissue particles. *Eur. J. Rheum. Inflam.* **1** (2), 130.

DOHERTY, M., WHICHER, J. and DIEPPE, P.A. (1982) Activation of the alternate pathway of complement by inflammatory particles *Ann. Rheum. Dis.* (In Press).

GLATT, M., DIEPPE, P. and WILLOUGHBY, D.A. (1979) Crystal-induced inflammation, enzyme release and the effect of drugs in the rat pleural space *J. Rheum.,* **6,** 251.

HASSELBACHER, P. (1979) C3 activation by monosodium urate monohydrate and other crystalline material. *Arth. Rheum.* **22,** 571.

KOZIN, F. and McCARTY, D.J. (1976) Protein absorption to monosodium urate, calcium pyrophosphate dihydrate and silica crystals. *Arth. Rheum.* **19,** 433.

MANDEL, N.S. (1976) The structural basis of membranolysis. *Ibid,* **19,** 439.

McCARTY, D.J. *et al.* (1981) 'Milwaukee Shoulder': association of micro-spheroids containing hydroxyapatite crystals, active collagenase, and neutral protease with rotator cuff defects. *Arth. Rheum.* **24,** 464.

PHELPS, P. and McCARTY, D.J. (1966) Crystal-induced inflammation in canine joints: Importance of polymorphonuclear leucocytes. *J. Exp. Med.* **124,** 115.

RAJAN, K.T. (1975) Observations on phagocytosis of urate crystals by polymorphonuclear leucocytes. *Ann. Rheum. Dis.* **34,** 54.

SPILBERG, I., GALLACHER, A. and MANDELL, B. (1973) Studies on crystal-induced chemotactic factor. *J. Lab. Clin. Med.* **85,** 631.

WALLINGFORD, W.R and MCCARTY, D.J. (1971) Differential membranolytic effects of sodium urate and calcium pyrophosphate dihydrate crystals. *J. Exp. Med.* **133,** 100.

WEISSMAN, G. (1971) The molecular basis of gout. *Hospital Practice* **6,** 43.

Chapter 6

PART TWO

GOUT

6.1 Introduction

Gout is the prototype of all crystal deposition diseases. Not only is it the oldest, and the disease with the finest history, but it also remains the best understood, easiest to diagnose and simplest to treat of all the conditions mentioned in this book. A good deal is known about the metabolism of uric acid and about the crystallization of its sodium salts; there is excellent evidence for the central role of the crystals in gouty arthritis. Nonetheless gout is not fully understood, and many remaining questions will be highlighted.

Gout is associated with the presence of crystals of monosodium urate monohydrate. The urate is a break-down product of the purine residues of nucleic acids, guanine and adenine. Fig. 6.1 is a simplified representation of some of the pathways involved in the metabolism of guanine and adenine, and the formation of urate (usually measured as its unionized form, uric acid). Uric acid is the metabolic end point, and as such is likely to be the major source of problems if there are imbalances of intake and excretion. The major part of this chapter will concentrate on the metabolism and the crystallization of urate, and the associated disease, gout. A number of other sparingly soluble compounds are formed from nucleic acids and these sometimes deposit in the tissues; they will also be discussed at the end of this chapter.

6.2 History

Gout has been known about since the time of ancient Egypt, and a urinary calculus containing urate salts has been recovered from a mummy 7000 years old. Gout was probably common amongst the ancient Greeks, and was well described by Hippocrates, whose three famous aphorisms are quoted in Table 6.1. The Romans also knew about the disease, and their physician Arelaeus was the first to suggest that a specific toxic substance might be the cause of the condition. Galen wrote voluminously about gout and was able to include the disease in his humoural theories; the inflammation and tophi

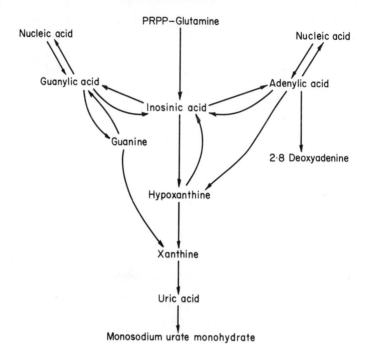

(PR PP = 5 Phosphoribosyl–1–Pyrophosphate)

Fig. 6.1 A simplified diagram of some of the pathways involved in purine metabolism.

Table 6.1   Three oft-quoted aphorisms of Hippocrates

Eunuchs do not take the gout (No. 28)
A woman does not take gout unless her menses be stopped (No. 29)
A young man does not take the gout until he indulges in coitus (No. 30)

being thought of as excretions of evil humours. There was little further advance in the understanding of gout until the time of Garrod, although excellent descriptions of the disease abound in early medical writings, and noteworthy landmarks are the introduction of hermodactyl (colchicine) by the Byzantine physician Alexander, and the wonderful descriptive writings of Thomas Sydenham (1624–1689), a London physician who suffered from gout himself.

The 18th and 19th Centuries were the heyday of gout. It was considered a social advantage to be a sufferer or to have gout in the family, and among many famous patients (Table 6.2) were royalty and politicians. Gouty attacks have undoubtedly influenced the course of history, including perhaps the Boston tea party, precipitated by Pitt during a particularly vicious attack. The noble ancestry of gout is impressive indeed, even Adam may have suffered! ('Our common ancestor Adam died of the gowte', anonymous monk *c.* 1450 A.D.). The impressive list of names shown in Table 6.2 has led Havelock Ellis and others to speculate that the gout is a marker of genius, and there remains evidence that I.Q. can be correlated with the levels of serum uric acid.

Table 6.2 Some famous gout sufferers (in no particular order!)

Charles Darwin	Edward Gibbon
Horace Walpole	Alexander Pope
Alfred Tennyson	Neville Chamberlain
Charles Scudmore	William Congreve
David Garrick	Thomas Graz
Queen Anne	Lord Chesterfield
W. S. Gilbert	W. H. Woollaston
Caleb Parry	Agrippa
William Harvey	George IV
John Hunter	Louis XIV
Samuel Johnson	Benjamin Franklin
Lord Palmerston	James I
Horatio Nelson	The Medici Brothers
Sir John Falstaff	Casanova
Harold the Saxon	Martin Luther
Oliver Cromwell	John Calvin
John Wesley	Lord Burghley
Cardinal Wolsey	Thomas Sydenham
Robert Boyle	Francis Bacon
Christopher Columbus	Isaac Newton
William Pitt	John Milton
Henry Fielding	Admiral Lord Howe

The word gout is derived from *gutta* meaning a drop, and was based on the humoural theory, which implied that the evil humours dropped down and spread out into the toe and were expressed as the disease. In 1797 William Hyde Woollaston (1766–1828) showed that the principal constituent of gouty tophi was a 'neutral compound consisting of lythic acid (uric acid)'. Scheele had found the same acid in some urinary stones, and the pioneer microscopist Van Leewenhoek discovered crystals in urine; this work suggested a chemical basis which could supercede the humoural theory of gout. However, for many years thereafter, majority opinion still believed that the urate salts were formed secondarily to the inflammation.

The modern history of gout, and of all crystal deposition diseases, began with the work of Sir Alfred Baring Garrod (1819–1907). In 1848 Garrod was a young assistant physician in University College Hospital, London. At that time he reported an extraction of urate of soda from the serum of gouty men, using the murexide test. Later Garrod introduced the now famous thread test, which demonstrated the precipitation of urate crystals on a thread passed through the serum of gouty sufferers, a test that has recently been recreated by Professor H. L. F. Currey in the London Hospital. Garrod was also an astute clinical observer of the gout, and presented his work in an excellent treatise on the subject. This set forth his views in a series of ten propositions (Table 6.3). Garrod's propositions were the first time that the precipitation of crystalline material in the body was suggested as the cause of the disease. His theory was not, however, widely accepted at that time, and although the association between high serum uric acid and gout was quickly established, the concept of crystals as a cause of disease only became respectable quite recently.

Table 6.3   The ten propositions of Sir Alfred Baring Garrod (1849) (Abridged)

First:	'In true gout uric acid is invariably present in the blood in abnormal quantities'
Second:	'. . . true gouty inflammation is always accompanied with deposition of urate of soda in the inflamed part'
Third:	'The deposit is crystalline and interstitial'
Fourth:	'The deposit may be looked upon as the cause, and not the effect of the gouty inflammation'
Fifth:	'The inflammation of the gout tends to the destruction of the urate of soda'
Sixth:	'The kidneys are implicated in gout'
Seventh:	'An impure state of the blood, arising principally from the presence of urate of soda, is the probable cause . . .'
Eighth:	'The causes . . . are such as lead to its (uric acid) retention in the blood'
Ninth:	'The causes exciting a gouty fit are those which induce a less alkaline condition . . . augment the formation of uric acid . . . or check the power of the kidneys for eliminating this principle'
Tenth:	'In no disease but true gout is there a deposition of urate of soda in the inflamed tissues'

In the latter part of the 19th Century and early 20th Century, the origin of uric acid from purines and the complex metabolic pathways involved were unravelled by a number of eminent chemists, including Emile Fischer. More recently inborn errors of metabolism which predispose to raised serum uric acid levels have been discovered.

Two German workers, Freudweilher and Hiss (of the Hiss Bundle) were working on crystals as a cause of inflammation at the turn of this century; they showed that uric acid and other powders were inert, whereas the crystalline sodium biurate and a number of other crystals caused intense inflammation in animal skin. This work was largely unknown until the early 1960s, when the introduction of polarized light microscopy for the study of synovial fluids marked the next great step forward in the story of crystal deposition disease. In 1961 McCarty and Hollander reported that crystals could be seen in nearly all gouty synovial fluids, but in no other condition. Shortly thereafter, they and other workers repeated and extended the work of Freudweilher and Hiss and showed that inflammation similar to that of the gout could be produced by injection of sodium biurate crystals. The combination of the presence of the crystals in all cases, and their ability to reproduce the inflammation, led to the simple concept of the crystals as the cause of gout being quickly accepted. Garrod was clearly some 110 years ahead of his time!

6.3   The metabolism of uric acid

In simple organisms, uric acid is a halfway stage in the total degradation to ammonia and carbon dioxide of purines, the adenine and guanine derivatives found in the nucleic acids, DNA and RNA. Many higher animals have a further degradative step beyond uric acid to form allantoin via the enzyme uricase. However, this enzyme has been lost in man, New World monkeys, birds and reptiles, and the pathways therefore end with the production of uric acid. The purines arise from the diet (exogenous) and from degradation of nucleic acid (endogenous). In addition, purines are synthesized via inosinic acid by a complex metabolic pathway, finely controlled by feedback inhibition. Input to the pool of uric acid in the body is therefore dependent on the balance between the diet, degradation and synthesis. The major pathways of excretion of uric acid are the urine and the faeces. About two-thirds (about 600 mg/day, or 5–10 % of the filtered load) is excreted in the urine, following filtration, partial reabsorption and re-excretion in the renal tubules. The remaining one-third is excreted via the gastro-intestinal tract, where it is degraded by bacterial action into ammonia and carbon dioxide. These basic pathways are shown in Fig. 6.2.

In a normal person, the total pool of uric acid amounts to about 1 g. In people with gout this may be elevated by two or three times, and in

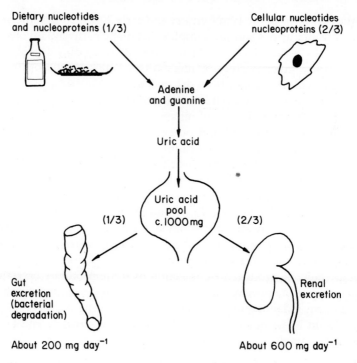

Fig. 6.2 Diagram of input and output of the body's pool of uric acid.

tophaceous gout may be as high as forty or fifty times normal. Normal plasma uric acid levels range from 100–400 μM (2–7 mg/100 ml) and tend to be higher in men than in women. Figs 6.3 and 6.4 show the normal distribution curve of uric acid levels in the community, and the relationship

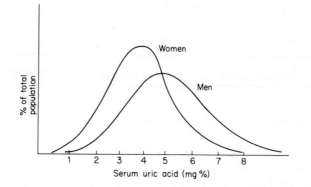

Fig. 6.3 Distribution of serum uric acid levels in the population.

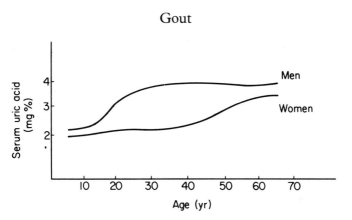

Fig. 6.4   Serum uric acid levels and age.

between uric acid levels and age (a conversion scale between the two commonly used measurements of uric acid can be found in Appendix II). Uric acid is fairly equally distributed throughout all the body fluids, although levels are rather lower in the saliva than elsewhere, and very low in cerebro-spinal fluid and sweat.

A key compound in the production of uric acid de novo is 5-phosphoribosyl-1-pyrophosphate (PRPP). This substance is formed via a synthetic enzyme (PRPP synthetase) and combines with glutamine to form a compound which is then degraded to inosinic acid. Inosinic acid, in turn, forms guanilic and adenilic acid, the purines which are incorporated into nucleic acids. This synthetic pathway is controlled by feedback inhibition, dependent on the levels of guanylic acid and adenylic acid. The acids can in turn form the bases guanine and adenine, but a key step is the so-called 'salvage pathways' that allow guanine and adenine to reform into the acidic forms to be incorporated into nucleic acid. Inosinic acid is broken down through inosine to hypoxanthine, xanthine and uric acid; but again, an important retrieval system allows hypoxanthine to be reformed to inosinic acid. These pathways have been outlined in Fig. 6.1.

Over the last three decades, identification of specific enzyme defects in the pathways of uric acid metabolism, leading to hyperuricaemia, have been described. The most important of these are outlined in a simplified form in Fig. 6.5. They are (1) deficiency of the enzyme hypoxanthine guanine phosphoribosyl transferase (HGPRT) leading to reduced salvage pathways for hypoxanthine and guanine, and thus an increase in the breakdown of inosinic acid and guanilic acid to form uric acid; (2) increased phosphoribosyl pyrophsphate synthetase (PRPP synthetase) leading to an excess of the basic component PRPP and overactivity of the whole pathway; (3) deficiency of adenine phosphoribosyl transferase (APRT) leading to reduced salvage of adenine to adenylic acid (Holmes *et al.*, 1975; Seegmiller, 1980).

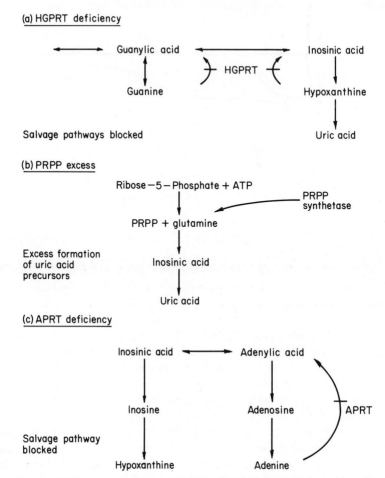

(a) HGPRT deficiency

Salvage pathways blocked

(b) PRPP excess

Excess formation
of uric acid
precursors

(c) APRT deficiency

Salvage pathway
blocked

Fig. 6.5   Three enzyme abnormalities that can cause hyperuricaemia.

6.4   Hyperuricaemia

The causes of hyperuricaemia are outlined in Figs 6.6 and 6.7 and Table 6.4.
High serum levels may result from a high purine diet; from increased tissue
turnover; from excess de novo synthesis of urate; or from reduced excretion
by the kidney or gut. However, as illustrated in Fig. 6.7, a combination of
two or more of these factors are often responsible for hyperuricaemia rather
than one alone. In the past a great deal of argument has been centred on the
number of hyperuricaemic individuals that are over-producers of uric acid,
and how many are under-secretors. Present evidence suggests that dietary
factors, including alcohol, are present in as many as 50%, that some 30% are

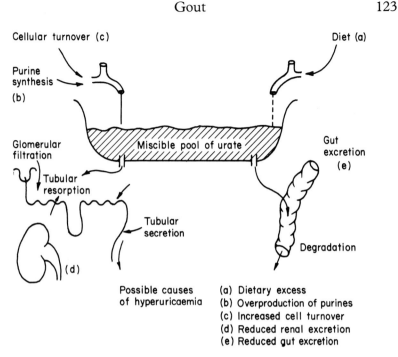

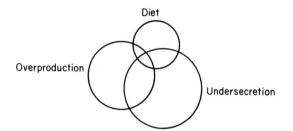

Fig. 6.6 Possible causes of hyperuricaemia.

Fig. 6.7 Interaction of the different causes of hyperuricaemia.

natural over-producers of uric-acid, and 70% tend to renal under-secretion. It is clear that a combination will operate in most people.

The associations of hyperuricaemia have been called 'the associations of plenty'. Hyperuricaemic individuals tend to have higher social class and high I.Q., to be over-weight and to drink more alcohol than others, as well as being hypertensive and prone to cardio-vascular disease (Table 6.5). Dietary factors certainly play a part in hyperuricaemia, although the relationship between diet, obesity, hypertriglyceridaemia, alcohol and hyperuricaemia is a complex one that is not fully understood. Meat typically

Table 6.4 Possible causes of hyperuricaemia

(1) Increased intake of purines in diet

(2) Overproduction of uric acid

    Primary:
    Idiopathic
    HGPRT deficiency
    APRT deficiency
    PRPP excess
    G. 6-PD deficiency

    Secondary:
    Myeloproliferative diseases
    Haemolysis
    Carcinomatosis
    Gaucher's disease

(3) Renal undersecretion of uric acid

    Idiopathic
    Secondary to:
    Chronic renal failure
    Hyperparathyroidism
    Lactic acidosis
    Starvation
    Lead poisoning
    Diuretics and other drugs
    Down's syndrome
    Myxoedema
    G. 6-PD deficiency
    Beryliosis

(4) Gut undersecretion of uric acid

Table 6.5 The associations of hyperuricaemia
'The associations of plenty'

High social class
High I.Q.
Alcohol excess
Obesity
Hypertriglyceridaemia
Hypertension
Cardiovascular disease

contains around 50 mg of purine nitrogen (equivalent to 170 mg uric acid) in every 100 g. Levels in other foods are much smaller. The normal diet would, therefore, only account for about 30% of excreted uric acid. Subjects fed large quantities of RNA (equivalent to about 1.2 kg of meat daily) doubled their serum uric acids. Similarly, changing from a normal to a purine free diet results in a 10–20% decrease in serum uric acid levels. However, alterations in the amount of meat eaten are not generally a major factor in hyperuricaemia, and it is important to know that a reducing diet, causing weight loss without a significant reduction in the purine content, will also lead to a 10–20% reduction in serum uric levels. Whether this is due to the relative increase in fluid content accompanying weight loss, or to a more complex metabolic interrelationship, is not yet clear. Excess alcohol probably induces hyperuricaemia by increasing de novo synthesis of urate, and not through calories or purines; however, very old port has been found to contain a high purine and lead content, and may have been important in Georgian England.

The causes of over-production of uric acid are listed in Table 6.4. Many hyperuricaemic individuals apparently have a general tendency to over-production of uric acid without there being any obvious enzyme defects. In between 30 and 50% of cases there is a family history, and presumably the tendency to over-produce is inherited. As already outlined, others have a specific genetically determined enzyme defect to explain their over-production. Secondary over-production of uric acid results from increased turnover of purines. A large number of haematological and neoplastic diseases can result in this, accounting for some 10% of gouty subjects. The majority are myeloproliferative diseases, in particular polycythaemia rubra vera.

The renal excretion of uric acid is complicated. Uric acid is filtered through the glomerulus, and then reabsorbed in the proximal tubule and resecreted back into the filtrate.

By using drugs such as pyrazinamide, which block tubular secretion, renal handling of urate has been extensively investigated. The present evidence favours a four-component system as illustrated in Fig. 6.8. Tubular secretion occurs in response to plasma uric acid concentration up to a maximum ($t_{max\,uric\,acid}$), and renal handling is, therefore, dependent on the number of functioning nephrons. Serum urate will rise in chronic renal failure when the glomerular filtration falls below about 20 ml min^{-1}. Many hyperuricaemic individuals seem to have abnormally low output of urinary urate in response to elevated serum levels, in the absence of significant renal failure. These patients also tend to secrete a more acid urine, and have a deficit of ammonium excretion. However, the exact tubular defect of gouty 'undersecretors' remains unclear.

A normal kidney has a remarkable capacity to increase urate clearance in

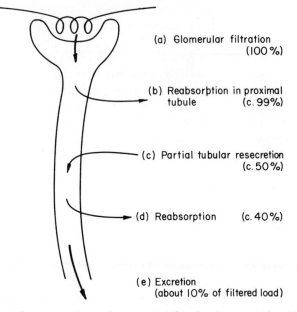

(a) Glomerular filtration
(100 %)

(b) Reabsorption in proximal
tubule (c. 99%)

(c) Partial tubular resecretion
(c. 50%)

(d) Reabsorption (c. 40%)

(e) Excretion
(about 10% of filtered load)

Fig. 6.8 Renal handling of uric acid (after Levinson *et al.*, 1980).

the face of an increased serum uric acid level, but a large number of drugs, renal diseases, and metabolic abnormalities such as lactic acidosis and hyperparathyroidism, can also result in reduced renal clearance of uric acid, and consequent hyperuricaemia (Table 6.4). Kidney disease in gout is discussed further in Chapter 10.

The excretion of uric acid in the faeces has not been investigated so thoroughly, although it has recently been shown to be subject to diurnal variation. Serum uric acid levels also fluctuate considerably from day to day, and vary according to diet, exercise and body rhythms.

6.5 Crystallization of urate

In keeping with our discussion of crystal growth in Chapter 3, we can approach the relationship between urate levels in the body and gout in a number of stages. The first and most important question is whether there are regions of the body which are continuously supersaturated with respect to monosodium urate. Having established this we will consider whether the supersaturation is high enough for spontaneous precipitation or whether the system will be metastable (see Fig. 3.6) in which case we will be concerned with the origins of the nuclei. We should also consider whether crystal growth is sufficiently rapid to occur during an attack.

As discussed above, urate is a general metabolic product whose concentration is uniform throughout most body fluids so that we need only consider the serum levels, which are easy to measure. The solubility of monosodium urate at pH 7.4, 37° C and 0.14 M sodium ions, i.e. under physiological conditions, is about 9.5 mg/100 ml or 0.57 mM. There is a small amount of urate binding to serum albumin and globulins which would raise the solubility. Osmotic effects of the dissolved proteins apparently reduce this to the measured solubility in serum of about 7 mg/100 ml, 0.42 mM. This correlates well with the risk of getting gout which becomes significant at serum uric acid levels around 7 mg/100 ml as shown in Table 6.6 from the Framingham Heart Study (Hall *et al.*, 1967). Thus hyperuricaemia, defined as a serum uric acid level above 7 mg/100 ml can be regarded as a necessary condition for gout. This corresponds to a requirement that the serum, and the rest of the body in equilibrium with the serum, be supersaturated with respect to monosodium urate.

Table 6.6 Relationship between serum uric acid levels and gout

Serum uric acid level	% risk of developing gout	
(mg/100 ml)	Men	Women
<6	0.9	0.15
6–6.9	2.8	4.6
7–7.9	17.3	26.1
8–8.9	27.5	—
>9	90	—

Reproduced with permission from the Framingham Heart Study, Hall *et al.*, 1967.

A few patients have been described who have had acute attacks of gout with uric acid levels below 7 mg/100 ml. However, their levels were probably above this at the time at which crystals were deposited in the connective tissues, rather than at the time of release into the joint space, triggering the attack.

The risk of developing gout increases dramatically with the degree of supersaturation, as one would expect of a disease initiated by crystal growth (Table 6.6).

Although the serum urate levels of gouty patients are above saturation they are still far below those at which precipitation occurs spontaneously *in vitro*, solutions of 80 mg/100 ml under physiological conditions have been reported as stable for 6 months. Thus in terms of Fig. 3.6, gouty patients are in the metastable region but close to the saturation line. Crystal nucleation will be a rare event arising from some special circumstances. This explains why some hyperuricaemic individuals do not get gout even at very high serum urate levels, and why several years may elapse between the detection of hyperuricaemia and the onset of the disease.

There have been many hypotheses regarding nucleation of urates, often arising from a knowledge of the associations and localization of gout attacks. Temperature is undoubtedly important, the solubility of urate decreases by half for a 10° C temperature decrease (Loeb, 1972). Thus a 9 mg/100 ml urate solution which is 30% supersaturated (c-s/s) at 37° C, is 230% supersaturated at 27° C, so that the tendency to nucleate will be much increased. This is probably a major factor in the peripheral distribution of gout attacks as superficial temperatures in the toes, fingers and ears can easily be 10° C below that of the body core; however the temperature *inside* the first metatarso-phalangeal joint (the major site of gout attacks) has not been accurately measured.

Crystals of monosodium urate are only precipitated in connective tissues. Within the joints the crystals can be found in cartilage, synovium and synovial fluid but cartilage is probably the principal area of deposition, at least initially. All the extra-articular sites of deposition and formation of tophi are also areas rich in connective tissue. Katz (1975) described a three-fold increase in serum uronic acid levels in patients with gout, and normalization of this abnormality with colchicine. It was suggested that this reflects increased turnover of connective tissue in patients with gout, and that this may be the inherited defect predisposing patients to crystal deposition. Katz and Schubert (1976) also showed that proteoglycan aggregates of connective tissue were capable of inhibiting urate precipitation and that they lost this property when degraded by a hyaluronidase. This suggested that tissue changes in cartilage might initiate precipitation. However, Perricone and Brandt (1978) have shown that this inhibition may be an artefact arising from high potassium levels in the proteoglycan preparation. There is as yet no evidence for significant interactions between urate and cartilage under physiological conditions. It may be that cartilage is one of the few regions where urate crystals can grow to maturity without being removed by macrophages before they are numerous and large enough to cause lysis.

Reduction of pH and addition of calcium ions have been shown to enhance precipitation but these effects do not seem likely to be important under physiological conditions (Wilcox and Khalaf, 1975). Gouty synovial fluid enhances precipitation from urate solutions *in vitro* (Tak *et al.*, 1980) but it is hard to ensure that this is not caused by small seed crystals of urate or phosphates in the fluid. Mechanical shock has also been shown to enhance nucleation. This is probably related to the effect known as 'secondary nucleation' in industrial crystallization where pre-existing crystals are chipped by collisions, producing an avalanche of small seeds. We have found that urate solutions that appear stable will rapidly crystallize if stirred or subjected to brief bouts of ultrasound. This process could be very active in the highly stressed and mobile environment of a joint. The first metacarpophalangeal joint is particularly highly stressed during walking.

Having been nucleated the crystals will grow as long as their surroundings remain supersaturated. Our own growth rate measurements at high urate concentrations show that the crystals grow at a rate which follows an exponential dependence on supersaturation (see Chapter 3). Extrapolation of this to a level equivalent to 9 mg/100 ml urate in serum gives a very slow growth rate of a small fraction of a micron per year. The true value may be higher than this but we have been unable to detect growth of seed crystals in synovial fluid and serum from gouty patients, suggesting that growth is indeed very slow. Lam Erwin and Nancollas (1981) also report very slow growth rates for urate crystals but suggest that the growth rates depend on (supersaturation)^2. They find that dissolution is very much faster than growth.

Thus in hyperuricaemia, urate is in metastable supersaturated solution from which some unknown and rare event causes nucleation, frequently in the joints of the extremities. These would be favoured by their low temperatures and by mechanical action and possibly by some special interaction between urate crystals and cartilage. Having nucleated, crystals grow slowly in the sheltered surroundings of the cartilage where the urate diffuses slowly in from the synovial fluid. Occasionally they are released into the fluid, when an inflammatory attack ensues. Fracture of the crystals in the fluid may lead to their multiplication.

6.6   The gout crystal: monosodium urate monohydrate

Crystals have been extracted from gouty tophi and synovial fluid, and examined by X-ray diffraction (Howell *et al.*, 1963). All investigators have found the same triclinic form of monosodium urate monohydrate, for which Mandel has determined the crystal structure (Fig. 6.9). The crystals consist of urate anions stacked in parallel sheets, interspersed with sodium ions which bond to four neighbouring urate anions via the oxygen atoms. The water molecules form hydrogen bonds with the purine rings. Recent studies in our own laboratory have shown that the crystals possess a net negative charge, with an electrophoretic mobility of about $1 \mu m \ s^{-1} \ V^{-1}$ cm, in physiological solution. This is consistent with the model of Mandel (Mandel and Mandel, 1976), which shows that net negatively charged oxygen atoms are prominent on the crystal surfaces. This also means that the crystal

Uric acid

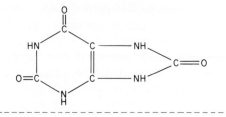

Monosodium urate monohydrate

$$NaC_5H_3N_4O_3 \cdot H_2O$$

Crystal structure (after Mandel and Mandel, 1976)

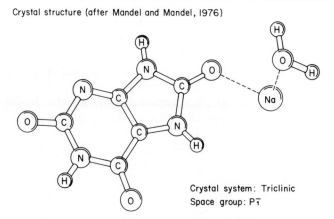

Crystal system: Triclinic
Space group: P$\bar{1}$

Fig. 6.9 The structure of uric acid and monosodium urate monohydrate.

surfaces provide a number of active charged groups, capable of attracting, and forming bonds with proteins and other biological material.

The morphology of the crystals has been studied by both light microscopy and transmission and scanning electron microscopy. The pioneering work of McCarty and Hollander in 1961 showed that the crystals could be identified by use of a polarized light microscope in the synovial fluid from virtually all cases of gout, and that they were found in no other disease. Numerous further studies have confirmed these findings, although occasional cases of well-established gout are apparently negative on examination by the light microscope. A likely explanation for this is that the crystals are occasionally just too small to be identified by this technique. However, in the majority of cases needle-shaped crystals of from 2 to 20 μm long and 0.5 to 2 μm wide, are seen. They are generally regular in size and shape in any one fluid, and very plentiful (Fig. 6.10–see colour Plate 3). When the fluid is extracted during an acute attack of gout, many of the crystals are seen within the leucocytes. They exhibit strong negative birefringence. Crystals extracted from tophi look very similar, although they tend to be somewhat larger and to form sheets or clumps which are enmeshed in an amorphous matrix.

In the electron microscope Pritzker and his colleagues (1978) have seen inter-connecting electron-lucent networks which differ between crystals from tophi and synovial fluid. However, this structure develops during exposure in the microscope and is a product of damage by the electron beam. It is not at all clear to what extent this reflects real internal structure. Electron diffraction patterns would be needed to show that the crystal structure was not being destroyed. The surfaces of the crystals are also different from the two sites. Crystals found within articular cartilage or tophi have a smooth surface, whereas those found in synovial fluid may have attached granular material that may be protein (Fig. 6.11). Fiechtner and Simkin (1981) have

Fig. 6.11   A crystal being phagocytosed: note the granular surface of the crystal
when viewed in the scanning electron microscope; this is probably due to
protein attached to the surface (× 120 000).

drawn attention to the presence of spherical aggregates of urate crystals in synovial fluid. Similar 'spherulites' are also formed at very high super-saturations *in vitro* but these also could arise through crystal fracture and multiplication in the cartilage.

Kozin and McCarty (1976) have shown that urate crystals are avid binders of protein, and Schumacher (1977) and his colleagues have demonstrated the presence of immunoglobulins, using immunofluorescence. The surface of the crystal has a net negative charge, and will tend to attract large polymeric compounds and proteins (see Chapter 5). Present evidence suggests that IgG is preferentially bound to the crystal surface, which might be expected as this is one of the least negatively charged of the serum proteins. Further, the Fab fragment of IgG has the lowest negative charge, and may bind preferentially, exposing the Fc fragment to interact with cell membranes (see Chapter 5).

6.7 The pathology of gout

It has already been mentioned that urate only crystallizes in connective tissues. The main sites are articular cartilage, periarticular soft tissues, bursae, epiphyseal bone and the kidneys. Tophi also occur on the ear, on the olecranon and patella bursae, and in tendon sheaths, and less commonly in a number of other sites.

Acute gout is characterized by the presence of numerous monosodium urate monohydrate crystals within the synovial fluid, often within the leucocytes, although there is no apparent relationship between the number of ingested crystals and the severity of the attack. The synovium shows a classical acute inflammatory reaction with large numbers of polymorphs infiltrating the membrane.

The pathology of chronic gout and of tophi is somewhat different. A classical tophus contains large numbers of crystals, often radially arranged, within a matrix, the nature of which is not fully certain. It appears to contain lipids and glycosaminoglycans, as well as small amounts of immunoglo-bulin. The tophi often grow radially, and are also characterized by a variable chronic inflammatory response, forming a rim of chronic inflammatory cells, giant cells and fibrosis around the periphery. The crystals are not ingested, and the factors activating continuing growth and the foreign body reaction are not clear. The joints in chronic gout often contain large deposits of urate on the surface of cartilage, in the synovium, and in the cartilage itself. The tophi also tend to extend into epiphyseal bone. The large deposits may be surrounded by chronic inflammatory tissue, which may form a pannus on the surface of the cartilage. Associated with extension of the tophus into the subchondral bone there is collapse and resorption of the trabeculae leaving a space characterized as the punched-out lesion seen in pathological samples, and on X-ray examination. Again, it is not clear how the growing tophus leads to formation of chronic soft tissue pannus and to destruction of bone. Other pathological features of chronic gout may include premature cardio-vascular changes, as well as renal disease, with or without deposition of urates. The deposition of urates and other purines in the kidney will be described further in Chapter 10.

6.8   Clinical aspects: acute and chronic gout

In its most classical form, the clinical picture of acute gout is characteristic and unmistakable. It was, therefore, recognized and categorized as a distinctive disease from the times of ancient Greece, long before there was any knowledge of uric acid or of crystals. However, gout is not always that easy to diagnose, nor is it easy to define. It is generally assumed that the clinical features of gout are a sequel to the formation of crystals, secondary to hyperuricaemia, and that the crystals are the cause of the clinical condition. This allows one to define gout on the basis of the presence of crystals in joints. The only problem with such a rigid definition would arise if crystallization was secondary to the appearance of joint inflammation. Although this suggestion has been put forward, the fact that the injection of crystals into synovial joints can reproduce an acute attack suggests that gout is indeed a crystal deposition disease, and that the crystals do cause the damage.

Most people with gout, if untreated, will probably go through the sequence of changes illustrated in Fig. 6.12. It seems likely that a long period of hyperuricaemia precedes any symptomatic problems in the majority of cases. Some event then triggers a first attack of gout, which may be followed after a period of months or years by further attacks. The period between these is known as the inter-critical phase of gout. Finally, if the

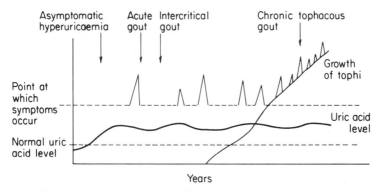

Fig. 6.12   A schematic representation of the development of the different stages of gout.

hyperuricaemia persists for a long enough time, and at a high enough level, larger deposits accumulate in joints, subcutaneous tissues and elsewhere (tophi). After reaching a certain size, which will obviously vary according to site, this may become symptomatic. This chronic tophaceous stage of gout tends to be accompanied by acute attacks, but is now rarely seen as it can be avoided with adequate treatment. There are, therefore, four stages of gout: (1) asymptomatic hyperuricaemia; (2) acute gout; (3) inter-critical gout; (4) chronic tophaceous gout.

The *acute attack* has been well described by many physicians over many years. Some of the best descriptions come from people like Thomas Sydenham (1624–1689) who suffered himself, and described the typical attack as follows:

'The victim goes to bed and sleeps in good health. About two o'clock in the morning he is awakened by a severe pain in the great toe, more rarely in the heel, ankle or instep. This pain is like that of a dislocation, and yet the parts feel as if cold water were poured over them. Then follow chills and shivers and a little fever. The pain, which was at first moderate, becomes more intense. With its intensity, the chills and shivers increase. After a time, this comes to its height, accommodating itself to the bones and ligaments of the tarsus and metatarsus. Now it is a violent stretching and tearing of the ligaments. Now it is a gnawing pain, and now a pressure and a tightening. So exquisite and lively, meanwhile, is the feeling of the part affected, that it cannot bear the weight of the bed clothes nor the jar of a person walking in the room. The night is passed in torture, sleeplessness, turning of the part affected, and perpetual change of posture.'

This vivid description, written in 1683, brings out all the characteristic features of the acute attack. They tend to occur in an otherwise fit person although some prominence has been given to a tendency to prodromal symptoms of general malaise, nervousness or irritability, occurring for a period of hours or days prior to an attack, and some sufferers claim that they are able to predict attacks. However, in the majority of cases, the attack starts fairly suddenly, and without warning. As described by Sydenham, the attack occurs more commonly at night than in the day, and the patient being wakened by pain is characteristic. Another feature is the very rapid onset. Few other joint diseases become so severe so quickly, and if a joint becomes maximally inflamed within a few hours of becoming uncomfortable, gout can be diagnosed with some confidence. The chills, shivers and fevers described by Sydenham, reflect the general systemic response to the severe local inflammation, reflected by leucocytosis, a mild fever, as well as general malaise in many cases (Fig. 6.13). The affected part becomes exquisitely painful and tender, more so probably than in any other form of joint disease. Severe reddening of the part is also characteristic, indicating increased

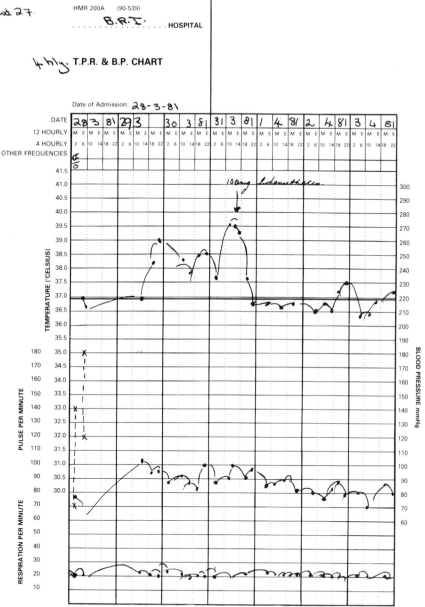

Fig. 6.13 Fever chart from a 46 year-old male patient with polyarticular gout, treated with indomethacin. The pyrexia was associated with leucytosis and led to an erroneous diagnosis of septic arthritis. Note the rapid resolution of fever and tachycardia after starting indomethacin.

vascularity and some inflammation extending into the periarticular and subcutaneous areas, as well as the joint itself. This may be followed by desquamation of the skin. Left untreated, the typical attack would slowly subside, although it may take 10 days to 3 weeks to do so. After that, the patient will once again be entirely normal, until the next attack occurs. In about 10% of cases attacks are polyarticular, affecting more than one joint, or 'spread' affecting first one joint and then another; this may cause diagnostic problems. The time between attacks is unpredictable. Patients may have one attack in their lifetime, or may have two or three attacks, separated by only a few weeks or months, followed by a long free period. The majority of patients will have one attack within a year of the last, the mean interval for an inter-critical period being some six to nine months. Although the condition is likely to be more severe the higher the uric acid concentration, it is hard to predict the progression of the acute attacks, or the development of the chronic tophaceous stage. The relationship between the duration and development of hyperuricaemia, and the development of tophi, is, however, statistically clear, as shown in Table 6.7.

Table 6.7 Relationship between tophi and the duration and severity of hyperuricaemia (after Gutman and Yu, 1971)

Years after first acute attack	% with tophi
5	36
10	55
20	65
30	70

Tophi	Mean serum uric acid level (mg/100 ml)
None	8.9 ± 1.5
Mild	9.9 ± 1.1
Moderate	9.9 ± 0.7
Severe	11.2 ± 2.0

Chronic tophaceous gout is defined by the presence of visible deposits of monosodium urate monohydrate. These accumulate slowly, and strangely, are often asymptomatic. The commonest sites are around the first metatarso-phalangeal joint, where most acute attacks occur, on the ears, in the olecranon bursa, on the elbow, and in the subcutaneous tissues of the hands (Fig. 6.14 and colour Plate 4). Deposits also commonly occur in

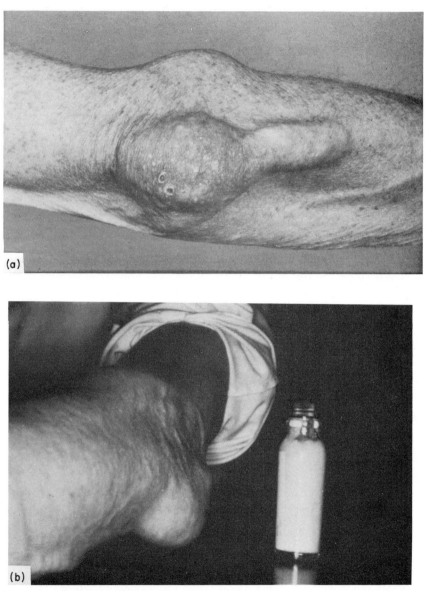

Fig. 6.14   (a) Chronic tophaceous gout showing tophi in the olecranon bursa, and along the ulnar border. Note the white skin deposits of urate crystals. (b) Olecranon bursitis in a case of gout which was misdiagnosed as rheumatoid disease. The fluid extracted from the bursa is shown; its white colour was due to the masses of urate crystals. (c) See colour Plate 4.

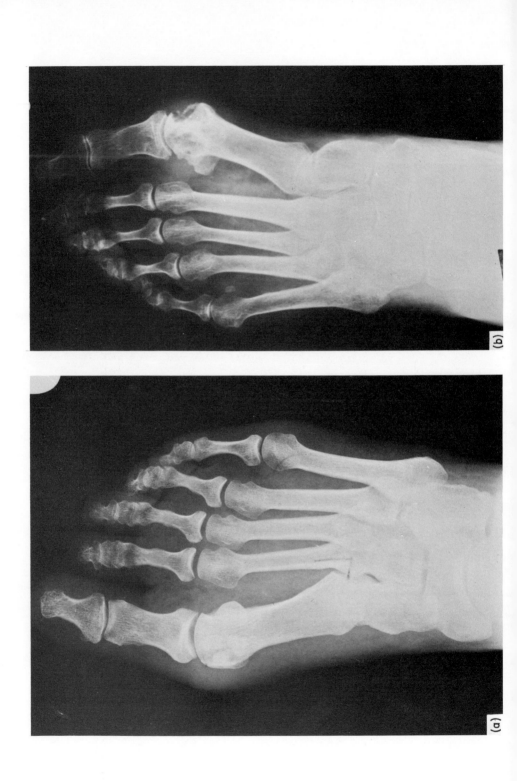

tendon sheaths, particularly around the achilles tendon and again in the hands. Less common sites include the eyes and eye lids, the penis (Paget's penile podagra'), tongue, lungs, intestine, and other viscera. Tophi are characteristically absent from voluntary muscle, liver, spleen and nervous tissue, where there is little connective tissue ground substance. The chronic stage of gout is usually accompanied by tophi occurring in articular cartilage and sub-articular bone, leading to a characteristic destructive arthropathy with radiological tophi and punched-out erosions (Fig. 6.15). The tophi often accumulate secondary deposits of hydroxyapatite as time goes on, leading to the appearance of flecks of calcific density on the X-rays (Fig. 6.16).

Patients with chronic gout also get secondary changes in the joints resembling osteoarthritis, with loss of articular cartilage and sclerosis of the underlying bone. This may result in chronic pain and stiffness in many joints.

The chronic tophaceous stage of gout is thus associated with a polyarticular destructive arthritis, with bone erosions, and tophi in many of the same sites as rheumatoid nodules. Not surprisingly it can be mis-diagnosed as rheumatoid arthritis, especially as acute attacks may subside or be absent in this phase of the disease.

Chronic gout may be complicated by renal failure or urinary calculi (Chapter 10), and by cardiovascular disease. The tophi themselves may cause problems, discharging through the skin, or becoming secondarily infected. Severe chronic gout is rarely seen now because of the efficacy of treatment (Chapter 11), but can be an extremely painful, disabling and unsightly disease, with extensive bone destruction (Fig. 6.17).

The pathogenesis of acute and chronic gout is only partially understood. As explained in Chapter 5, much research has been done on the inflammatory potential of crystals of monosodium urate monohydrate, and there can be little doubt that the crystals cause the inflammation. Why the attacks start and stop is less clear (see Section 6.11). The chronic destructive lesions that occur in the bone and cartilage in tophaceous gout are not understood. The mechanical pressure and abrasion caused by the tophi hardly seem to be an adequate explanation of the characteristic bone erosions. Whether the crystals can activate bone resorption by some other mechanism remains to be seen.

Fig. 6.15 Radiographs of gout of the first metatarsophalangeal joint. (a) Acute gout: the radiograph shows the soft tissue swelling, which is the only abnormality seen in a first attack of gout. (b) Chronic gout: the typical para-articular bone erosions and secondary osteoarthritis of chronic gout of the great toe are shown.

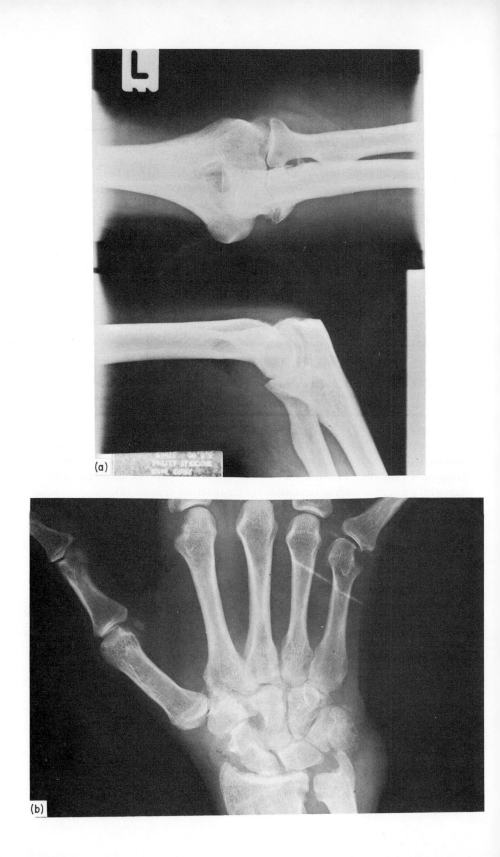

6.9 Epidemiology of gout

Who gets gout? There are a number of large surveys of patients with gout, and the data presented here mostly comes from the series of Grahame and Scott (1970), and Gutman and Yu (1971), which give very similar figures. It is principally a disease of men. Only 10% of sufferers are women, and those women that do get it tend to be older at the time of onset and to have a less severe disease and fewer renal problems. The mean age of onset is 40–45 yr in men and 55 in women.

Gout is usually divided into primary and secondary types. Most surveys show that the majority have primary gout, without any other disease causing the hyperuricaemia. They often have a family history of gout (36%), a body weight above 15% greater than the expected normal (48%), a history of excess alcohol consumption (37%), and hypertension (50%). Only about 10% of patients have secondary gout. Hospital-based series tend to show a predominance of myelo-proliferative diseases, in particular polycythemia rubra vera, in this group; in contrast general practitioner surveys show a predominance of diuretic therapy. In our own series diuretics are playing an increasing and important part, not only in asymptomatic hyperuricaemia but also in patients developing gout.

Patients in our gout clinic tend to fall into two categories. The first is middle-aged men, with a tendency to high alcohol consumption and obesity (Fig. 6.18). Most are undersecretors of uric acid without any other apparent causative factor. The second category of patient is the older person, often female, usually on diuretic therapy. These elderly ladies are particularly prone to tophi.

From what has been said about gout, one can construct a picture of a typical sufferer in England in the 18th Century (the hey-day of gout), and contrast it with one in the 20th Century.

6.9.1 The 18th Century gout sufferer

He would be middle-aged, obese and perhaps a member of the nobility, who typically enjoyed large quantities of food, strong beer, red wine, and port. He would be ravaged by frequent attacks of gout, usually affecting his toe, leaving him laid-up for several days with his foot on the gout stool, and attended by physicians and nurses, who would swathe the affected part in

Fig. 6.16 Gouty erosions. (a) A typical gouty erosion of the head of the radius. Note the hook of bone extending around the tophus, and the sclerotic margin of the erosion. (b) Gout of the wrist joint showing the large, well demarcated, sclerotic erosions in the carpal bones, and soft tissue swelling.

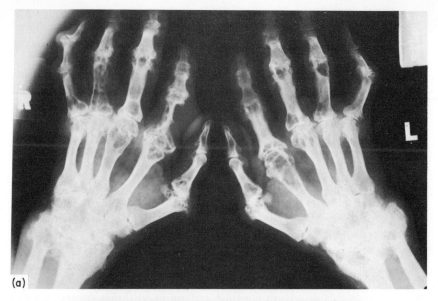

(a)

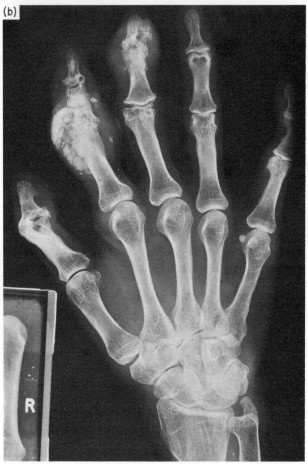

(b)

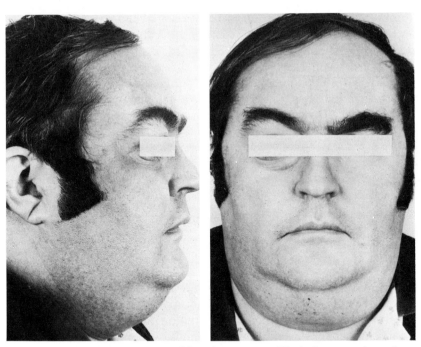

Fig. 6.18 A middle aged man with primary gout. He is hypertensive, obese,
hyperlipidaemic and partial to beer. He is an undersecretor of uric acid,
and has a family history of gout and ischaemic heart disease.

heavy bandages and bathe it with soothing lotions. As time goes on the
patient may develop tophi, which, as in the case of Horace Walpole, might
allow him to chalk his card board or dart score, but would also cause him to
be the butt for ribald jokes amongst his fellows. With a further passage of
time, dropsy might intervene, indicating renal failure, which would
terminate the illness.

6.9.2   The 20th Century gout sufferer

A mildly obese, hypertensive 'type A personality', the Managing Director of
a small company. He is awoken in the early hours by excruciating pain in his

---

Fig. 6.17 Radiographs of chronic tophaceous gout. (a) Severe chronic tophaceous
gout with extensive cystic changes and erosions in the bones of the hand,
due to compression and destruction of bone caused by the urate deposits.
(b) Calcified tophi in the hand, due to secondary deposition of
hydroxyapatite crystals in the urate tophus.

toe some months after being put on diuretics for hypertension by his general practitioner. A visit to his doctor confirms his fears that he has got the gout, and he protests, vehemently that he is not the drinker that he assumes the gout sufferer to be. His uric acid is measured, and found to be modestly elevated, and alteration in his anti-hypertensive therapy is all that is required. If he is unlucky, and has a familial tendency or greater predisposition to hyperuricaemia and gout, he may get further attacks, but these are adequately treated by hypouricaemic drugs, and tophi and renal disease are not a common feature of the 20th Century gout sufferer. He shares in common with his predecessor, though, his enjoyment of good living, high alcohol intake, elevated position in society, and the ribald jokes of his fellows.

6.10 Diagnosis

The diagnosis of gout can be considered at several different levels. At the clinical level the typical acute attack, as described, is highly characteristic. However, many attacks are minor, or atypical, and in some 10% of cases polyarticular attacks occur, causing diagnostic confusion. Other diseases can also mimic gout (pseudopodagra, Huskisson and Balme, 1972). Similarly, in the late tophaceous stages, the sites of distribution of tophi may be similar to those of rheumatoid and other nodules, and the differential diagnosis may include many other chronic rheumatic diseases.

At the radiological level, acute gout shows no specific features, but chronic gout is characterized by the soft tissue tophi and bony erosions. The erosions are usually round or oval, with their long axis parallel to the axis of the bone. They look 'punched-out', being well-defined, with sclerotic borders, and sometimes an overhanging margin (Martel's hooks, Martel, 1968). The larger an erosion, the more likely it is to be gouty. Tophi may calcify and occasionally ossify (Fig. 6.15).

Blood tests usually reveal hyperuricaemia, providing a guide to the likelihood of gout. However, as already explained, most people with hyperuricaemia do not get gout, and a few with normal serum uric acids do. The blood, then, is not a diagnostic test, and measurement of uric acid leads to the diagnosis of much more 'non-gout' than true gout.

The best method of diagnosis is by identification of urate crystals in synovial fluid using the polarized light microscope. Further absolute identification of them as urate, by dissolving with uricase, or by crystallographic methods, is not usually necessary. (A guide to the investigation of gout and hyperuricaemia, and the use of polarized light microscopy is given in Appendices I and II.) Treatment is outlined in Chapter 11.

6.11   Unanswered questions regarding gout

Many writings on gout, including this one, attempt to carry the reader through from an understanding of hyperuricaemia and the crystallization of urate, to the acute attack and tophi that develop as a consequence of the deposition of crystals. However, this is to suggest that the clinical features can be explained on the basis of the crystals and their metabolic background, and this is only a partial truth. There are a number of unexplained observations about the gout, some of which can now be discussed.

6.11.1   The distribution of gout

Table 6.8 shows the distribution of acute attacks of gout in the different joints of the body. The great toe is the site of 70% of all attacks, while the hip and the shoulder are spared in nearly all cases. The low temperature of the extremities may be important but is inadequate to explain this specificity. It has been asserted that the first metatarso-phalangeal joint is the most highly stressed in the body, and this may be important in leading to fracture and rapid multiplication of growing crystals. However, gout does not follow the distribution of osteoarthritis or any other joint disease, and no adequate explanation of its joint pattern exists.

6.11.2   What causes the acute attack?

Factors that might promote crystal nucleation and growth have already been

Table 6.8   Frequency of gout in different joint tissues. (Adapted from various sources)

	Joint sites of acute gout	
	First attack (%)	All attacks (%)
1st metatarso-phalangeal	53.4	77.5
Ankle	26.2	48.0
Tarsus	14.0	27.9
Other metatarso-phalangeal	7.6	14.5
Knee	24.9	53.1
Elbow	4.5	20.3
Carpus	6.3	19.5
Proximal inter-phalangeal joints	6.4	15.7
Metacarpo-phalanges	3.5	16.9
Distal inter-phalangeal joints	5.8	5.8
Hip	5.2	5.2
Shoulder	(very rare in all series)	

mentioned but none seem able to give sufficiently rapid growth to explain an acute attack. In clinical practice, many attacks are promoted by medical illness, drugs, surgery or other trauma (Table 6.9). However, as shown in the table, the theoretical and clinical actualities are difficult to marry together. Release of quantities of pre-existing crystals from the cartilage therefore seem to be the most likely cause of acute attacks.

Table 6.9   Possible factors precipitating acute attacks of gout

---

(1) Clinically recorded 'causes' of exacerbation of acute attacks of gout
(approximate figures from Gutman and Yu, 1971)

Severe medical illness	20%
Physical trauma	15%
Psychological trauma	15%
Excesses of food and drink	20%
Surgery	10%
Others	10%

(2) Proposed causes of growth of crystals in synovial fluid
     Physical trauma
     Alteration in temperature
     Alteration in pH
     Change in protein binding
     Resorption of extracellular fluid

(3) Causes of release of crystals into joint space
     Local fracture of cartilage by trauma
     Partial dissolution resulting in loosening of deposits

6.11.3   Why does the attack stop?

Left untreated, acute attacks of gout resolve spontaneously within a matter of a few days or weeks. It remains unclear why this should be, although a number of possible factors are suggested in Table 6.10. A combination of intrinsic anti-inflammatory activity and crystal removal seems most likely.

6.11.4   Why in the middle of the night?

The acute attack of gout occurs much more frequently at night than in the day, as any general practitioner, working in an affluent area, will agree. Simkin (1977) has suggested that this is due to the accumulation of fluid in joints on a day's use, followed by a period of resorption at night, at which time water is reabsorbed faster than uric acid, leading to a point at two o'clock in the morning, at which the uric acid levels are higher than elsewhere in the body fluids. In view of the slow growth of crystals this seems

Table 6.10 Possible causes for the self-limiting nature of the acute attack of gout

---

(1) Intrinsic anti-inflammatory activity:
 Anti-inflammatory acute-phase response
 Local anti-inflammatory cascade
 Anti-inflammatory prostaglandins

(2) Increased solubility of uric acid:
 Increased temperature via blood flow
 Altered pH
 Alteration in connective tissues or protein binding

(3) Removal of crystals:
 Phagocytic uptake and removal

(4) Alteration of crystal surface to render it less inflammatory:
 Alteration in protein coating

---

unlikely to be important, and this hypothesis remains to be substantiated. Other explanations such as alteration in pH and temperature, diurnal variations in urate metabolism or other chemical factors in the tissues, may also be relevant. However, if the attack is caused by crystal release rather than de novo crystallization the timing could also reflect a delayed response to a daytime event.

6.11.5 Why red desquamated skin?

Very few joint diseases cause reddening and changes of the overlying skin to the extent that occurs in gout. This could, perhaps, be explained merely by the intensity of the inflammation leading to marked hyperaemia. However, gout in the soft tissues and periarticular areas does occur, and can cause a condition like cellulitis. There may, therefore, be an active inflammatory process, occurring in the periarticular tissues as well as in the synovium, in most attacks.

6.11.6 Why aged 50?

As has been mentioned, levels of uric acid in the body reach their peak in men at the age of 20, but most people do not get gout until they are 40 or 50. This may be due to the need for other changes in connective tissue, before precipitation can occur, but this, too, remains unproven. Alternatively it may reflect very slow crystal nucleation and growth; thus it may take 20 years before a large enough deposit has formed to shed the dose of crystals required to excite an inflammatory reaction.

6.11.7   Why do some people have crystals and no gout?

Several investigators have described joints from which crystals can be extracted, but which remain without inflammation. Presumably in these cases there is some block to the factors discussed in Chapter 5 which trigger the acute inflammatory response; alternatively the number of crystals may be insufficient or their surfaces may be protected from causing damage by some surface absorption.

6.11.8   Why do so few people get secondary gout?

The data quoted on the relationship between hyperuricaemia and gout refers to 'primary' gout. Patients with myeloproliferative diseases, renal failure or neoplasia often develop very high uric acid levels, but relatively few get gout. Possibly a long duration of hyperuricaemia is needed, or perhaps they get crystals, but cannot respond with inflammation. Investigation of their joint fluid and synovial reactivity would be of interest.

6.11.9   Why are gout and rheumatoid arthritis mutually exclusive?

There are very few reports of gout and rheumatoid arthritis (R.A.) occurring in the same person. Several interesting hypotheses have been put forward to explain this, but none are convincing. In the adjuvant arthritis model in the rat, hyperuricaemia suppresses the disease, perhaps by immunosuppression. We have found that rheumatoid patients respond normally to intradermal injections of monosodium urate monohydrate crystals, but the ability of urate crystals to bind the abnormal immunoglobulins found in R.A. may be important.

6.11.10   Association with other crystal diseases

An association between gout and pyrophosphate deposition (Chapter 7) has been described, and apatite can be deposited in gouty tophi. These associations may be due to epitaxial crystal growth (Chapter 3); alternatively, the connective tissue defect postulated to predispose some patients to gout, may also enhance the formation of other crystal species.

6.11.11   Gout, hyperlipidaemia and cardiovascular disease

The association between gout and cardiovascular disease remains ill-understood. Many gouty patients are hyperlipidaemic (mainly type IV hyperlipidaemia, present in about 50%), but this appears to be independent of the hyperuricaemia. A deficiency in the lipoprotein fraction apo CII has recently been described by Macfarlane and colleagues (1981), and this may explain the elevated triglyceride levels; consequent reduction in HDL2, could perhaps explain a predisposition of myocardial infarction.

These, and many other questions, remain unanswered in gout, and it seems likely that this disease will remain a fascinating and instructive one to study for many years to come, and that it will be a long time before gout reveals all its secrets.

6.12   Other purine disorders associated with crystals

Of the various metabolic abnormalities of purine biosynthesis, only gout and hyperuricaemia are common. However, purine biosynthesis does lead to the production of a number of other relatively insoluble compounds, and occasional instances of their deposition have been described. In addition, brief mention will be made here of a rare gouty syndrome.

6.12.1   The Lesch–Nyhan Syndrome

As outlined above, total deficiency of the enzyme HGPRT leads to hyperuricaemia by loss of the salvage pathways of hypoxanthine and guanine. Partial deficiency of this enzyme may lead to premature severe gout, and, where there is a family history and severe gout in relatively young people, measurement of the enzyme (which can be made on erythrocytes) is worthwhile for genetic counselling. In its complete form, the enzyme deficiency leads to a characteristic clinical picture, which includes striking self-mutilation and abnormalities of the central nervous system, including choreoathetosis and spasticity. This, in conjunction with gout, leads to a distressing clinical picture of spastic children who may have to be restrained from causing themselves severe damage, which includes biting their lips and limbs, with loss of digits. The cause of the CNS disturbance is not clear (Kelley *et al.*, 1969).

6.12.2   Saturnine gout

Garrod noticed that many of his gouty patients had lead poisoning. Ball and Sorenson (1969) have more recently investigated several patients with 'saturnine gout' and found them to be very poor excretors of uric acid, suggesting that renal tubular damage causes the association. The condition is common in areas in which illegal alcoholic spirits are brewed in contact with lead ('moonshine gout'). As mentioned below, it has been suggested that guanine crystals might also be deposited in patients with lead poisoning.

6.12.3   Hereditary xanthinuria

This rare disorder results from deficiency of xanthine oxidase, so that hypoxanthine and xanthine accumulate instead of uric acid. In addition to the production of xanthine renal stones (see Chapter 10), other clinical manifestations can include a myopathy with deposition of xanthine and hypoxanthine crystals in the muscles. Xanthine crystals have also been described as being deposited in children treated for the Lesch–Nyhan Syndrome.

6.12.4   Hypoxanthine and oxypurinol

Crystals of hypoxanthine and oxypurinol, as well as microcrystals of xanthine, have been detected in muscles biopsied from patients treated with allopurinol for a long time. Allopurinol blocks the breakdown of xanthine to uric acid and produces an iatrogenic form of xanthinuria. However, although this can occasionally lead to crystals being deposited, this does not apparently lead to any clinical manifestations. It is interesting to note that xanthine, hypoxanthine and oxypurinol crystals have all been detected in the muscle, whereas muscle seems to be one of the few sites at which urate deposition and tophi are extremely rare.

6.12.5   Guanine

Guanine crystals have recently been identified in pig ephiphyseal cartilage following *in vivo* inhibition of guanase by lead, given to produce experimental lead poisoning. These experiments used high doses of lead, in a species of animal with different enzymic pathways to that of man, but it has led to the suggestion that guanine crystals might be a feature of saturnine gout.

6.12.6   2, 8 Dihydroxyadenine

Recently a small number of patients have been described with severe APRT deficiency in whom crystals of 2, 8 dihydroxyadenine, a breakdown product of adenylic acid, formed. This led to severe renal calculi, although deposition of this compound in the joints remains to be described (Simmonds *et al.*, 1976).

6.12.7   Defects of purine metabolism and immunodeficiency

A few rare enzyme defects in the pathways of purine metabolism are also associated with abnormal function of T or B cells, or both (Seegmiller, 1980). This interesting finding may provide new insights into the genetic control of immune responses, as well as giving gout a new dimension.

6.13 Summary and conclusions

Uric acid is the most important of several sparingly soluble products of the metabolism of purine derivatives.

Hyperuricaemia, defined as a serum level above the saturation of uric acid, occurs in about 5% of the adult male population. It may be due to excess ingestion, intrinsic overproduction or reduced renal secretion. Although much is known about the complex metabolic pathways involved, a specific enzymic or transport defect can only be detected in a minority of cases.

Gout is a disease which can be clearly related to the formation of monosodium urate crystals in the joints of hyperuricaemic individuals. The risk of gout is proportional to the degree of hyperuricaemia, but gout has a prevalence of less than 0.5%, and why crystallization occurs in some, but by no means all, hyperuricaemic people is not known.

The crystals are well characterized, and much is known about their ability to cause inflammation, and about the associated clinical phenomenon – acute gout. The destructive lesions of chronic articular gout are less well understood.

Gout has been called the 'king of diseases, and disease of kings', and has a splendid history. It is clinically distinct, and its association with supersaturation and crystallization of urate is well established. It is, therefore, the best characterized crystal deposition disease of joints. Nevertheless, many aspects of gout remain ill-understood, and relatively little is known about conditions associated with crystals of other purine derivatives. Gout is likely to remain of research interest for many years, although the present level of understanding has already led to effective treatment; a goal not yet attained for the diseases described in the next two chapters.

Further reading

COPEMAN, W.C. (1964) *A short history of the gout*, University of California Press, Berkeley, California.

FREUDWEILER, M. (1899) Studies on the nature of gouty tophi, in McCarty, D. J., *Ann. Int. Med.* (1964) **60**, 486–505.

GARROD, A.B. (1876) *A treatise on gout and rheumatic arthritis*, Longmans, Green and Co, London.

KELLEY, W.N. (ed) (1977) Crystal-induced arthropathies, in *Clin. Rheum. Dis.* **3**, 1.

SCHUMACHER, H.R. (1978) *Gout and Pseudogout*, Medical Examination Press, New York.

SCOTT, J.T. (1978) New knowledge of the pathogenesis of gout. *J. Clin. Path.* **12**, 205–13.

Text references

BALL, G.V. and SORENSEN, L.B. (1969) Pathogenesis of hyperuricaemia in saturnine gout. *New Eng. J. Med.* **280,** 1199.

FEICHTNER, J.J. and SIMKIN, P.A. (1981) Urate spherulites in gouty synovia. *JAMA* **245,** 1533.

GRAHAME, R. and SCOTT, J.T. (1970) Clinical survey of 354 patients with gout. *Ann. Rheum. Dis.* **29,** 461.

GUTMAN, A.B. and YU, F.T. (1971) *Gout, a clinical comprehensive,* Medcome, New York.

HALL, A.P., BARRY, P.E., DAWBER, T.R. and McNAMARA, P.M. (1967) Epidemiology of gout and hyperuricaemia. *Am. J. Med.* **42,** 27.

HOLMES, E.W., KELLEY, W.N. and WYNGAARDEN, J.B. (1975) Control of purine biosynthesis in normal and pathological states. *Bull. Rheum. Dis.* **26,** 848.

HOWELL, R.R., EVANS, E.D. and SEEGMILLER, J.E. (1963) X-ray diffraction studies of the tophaceous deposits in gout. *Arth. Rheum.* **6,** 97.

HUSKISSON, E.C. and BALME, H.W. (1972) Pseudopodagra. *Lancet* **ii,** 269.

KATZ, W.A. (1975) Deposition of urate crystals in gout: altered connective tissue metabolism. *Arth. Rheum.* **18,** 751.

KATZ, W.A. and SCHUBERT, M. (1976) The interaction of monosodium urate with connective tissue components. *J. Clin. Invest.* **49,** 1783.

KELLEY, W.N., GREENE, M.L., ROSENBLOOM, F.M., HENDERSON, J.F. and SEEGMILLER, J.E. (1969) Hypoxanthine-guanine phosphoribosyltransferase deficiency in gout. *Ann. Intern. Med.* **70,** 155.

KOZIN, F. and McCARTY, D.J. (1976) Protein absorption to monosodium urate, calcium pyrophosphate dihydrate and silica crystals. *Arth. Rheum.* **19,** 433.

LAM ERWIN, C-Y. and NANCOLLAS, G.H. (1981) The crystallization and dissolution of sodium urate. *J. Crystal Growth* **53,** 215.

LEVINSON, D.J. and SORENSON, L.B. (1980) Renal handling of uric acid in normal and gouty subjects: evidence for a 4-component system. *Ann. Rheum. Dis.* **39,** 173.

LOEB, J.N. (1972) The influence of temperature on the solubility of monosodium urate. *Arth. Rheum.* **15,** 189.

MANDEL, N.S. and MANDEL, G.S. (1976) Monosodium urate monohydrate: the gout culprit. *J. Am. Chem. Soc.* **98,** 2319.

MARTEL, W. (1968) Overhanging margin of bone: roentgenologic manifestation of gout. *Radiology* **91,** 755.

McCARTY, D.J. and HOLLANDER, J.L. (1961) Identification of urate crystals in gouty synovial fluid. *Ann. Intern. Med.* **54,** 452.

MACFARLANE, D.G., SLADE, R. and DIEPPE, P.A. (1981) Gout, lipoproteins and vascular disease. *Ann. Rheum. Dis.* **41,** 200.

PERRICONE, E. and BRANDT, K.D. (1978) Enhancement of urate solubility by connective tissue. *Arth. Rheum.* **21,** 453.

PRITZKER, K.P.H., ZAIN, C.E., NYBORG, S.C., LUK, S.C. and HOUPT, J.B. (1978) The ultra-structure of urate crystals in gout. *J. Rheumatol.* **5,** 7.

SEEGMILLER, J.E. (1980) Human aberrations of purine metabolism and their significance for rheumatology *Ann. Rheum. Dis.*, **39**, 103–117.

SIMKIN, P.A. (1977) The pathogenesis of podagra. *Ann. Intern. Med.* **86**, 230.

SIMMONDS, H.A., VAN ACKER, K. J., CAMERON, S., and SNEDDON, W. (1976) The identification of 2, 8-dihydroxyadenine, a new component of urinary stones. *Biochem. J.* **157**, 485–7.

SOKOLOFF, L. (1957) The pathology of gout. *Metabolism* **6**, 230–43.

SYDENHAM, T. (1683) *Tractatus de podagra et hydrope*, G. Kettelby, London.

TALBOTT, J.H. (1957) *Gout*, Grune and Stratton, New York.

TAK, H.K., COOPER, S.M. and WILCOX, W.R. (1980) Studies on the nucleation of monosodium urate at 37° C. *Arth. Rheum.* **23**, 574.

WILCOX, W.R. and KHALAF, A.A. (1975) Nucleation of monosodium urate crystals. *Ann. Rheum. Dis.* **34**, 332.

WYNGAARDEN, J.B. and KELLEY, W.N. (1976) *Gout and hyperuricaemia*, Grune and Stratton, New York.

Chapter 7

CALCIUM PYROPHOSPHATE DIHYDRATE DEPOSITION

7.1 Introduction

Excess urate production is an example of a generalized metabolic abnormality in which precipitation of the salt could occur anywhere; the site is decided by secondary local tissue factors predisposing to crystallization. By contrast, in most cases where calcium salts are deposited in the joints, the fault is a more localized one; there is either a local increase in metabolic activity, or loss of inhibition to crystal nucleation, such that the deposition occurs at the site of the primary abnormality (Fig. 7.1).

Using X-ray analysis a number of different crystals have been identified in calcified joint tissues, and extra-articular calcific deposits (Table 7.1). In knee meniscal cartilage, Gatter and McCarty (1967) found calcium pyrophosphate dihydrate, dicalcium phosphate dihydrate and hydroxy-apatite. By contrast, nearly all extra-articular calcification consisted of hydroxyapatite alone, although small amounts of calcium orthophosphate were also seen. Other calcium phosphates may also exist rarely or transiently in calcified areas.

The importance of these compounds in rheumatology has only recently been appreciated. Whereas gout has a history which stretches back to the earliest medical records, the diseases mentioned in this chapter have only been described during the last 20 years. Their discovery depended on the application of technical innovations unknown to Hippocrates. Radiography of the joints shows up the larger calcific deposits, and polarized light microscopy or electron microscopy of joint tissue and fluid allows the crystals to be seen and identified. The recent increase in the use of these techniques has been accompanied by a parallel increase in awareness of the frequency of calcium deposition in pathological joints. Calcium pyro-phosphate dihydrate will be considered in this chapter, and discussion of the causes and associations of articular and periarticular deposition of calcium orthophosphate salts will follow in Chapter 8.

154

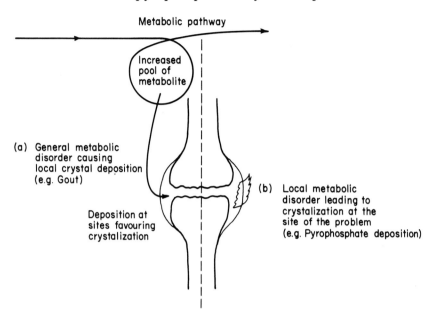

Fig. 7.1   Crystal deposition in joints: there are two types of crystal deposition
disease of the joints: (a) resulting from a general metabolic disease; and
(b) due to a local abnormality.

Table 7.1   Cadaveric knee joint fibrocartilage (215 cases)

	Frequency of calcific deposits (%)
(a) Calcium pyrophosphate dihydrate	3.2
(b) Dicalcium phosphate dihydrate	2.3
(c) Hydroxyapatite	1.4

Small unidentified calcified areas found in many other cases
Reproduced with permission from Gatter and McCarty, 1967.

7.2   Pyrophosphate deposition

The use of the polarized light microscope for systematic examination of
many synovial fluids was begun in the early 1960s, and quickly led to the
positive identification of urate crystals in gout. At the same time, a number
of patients with a gout-like syndrome were found to have other birefringent
crystals within their synovial fluids. These crystals were identified in 1962
by X-ray powder diffraction as calcium pyrophosphate dihydrate in triclinic
and monoclinic forms. The patients suffered from acute attacks resembling

gout, although tending to affect more proximal joints, and rather less severe. This phenomenon was initially termed 'pseudogout' (McCarty *et al.*, 1962).

A similar disease had been found in the 1950s amongst a small group of Hungarian families living in Slovakia (Zitnan and Sitaj, 1976). In these cases it was identified on the radiographic appearance of the knee joints, which showed small linear areas of calcification of the menisci and hyaline cartilage. A number of families were identified whose members were prone to attacks of acute inflammatory arthritis in the affected joints as well as to premature osteoarthritis. The radiographic appearance, which had been observed but poorly documented prior to this period, was due to the deposition of calcium salts in the cartilage, and is termed 'chondrocalcinosis'.

It was soon realized that patients with the radiographic appearance of chondrocalcinosis and those with pseudogout were suffering from the same condition, characterized by the deposition of monoclinic or triclinic calcium pyrophosphate dihydrate in joint cartilage. This led to a change in terminology, and to the application of terms such as 'pyrophosphate arthropathy' or 'calcium pyrophosphate dihydrate crystal deposition disease'. The recent discovery that a range of compounds can be involved in calcified cartilage, and the fact that they cannot always be identified either by polarized light microscopy or by radiography is potentially confusing. We would suggest the use of a strict nomenclature in dealing with these phenomena. The suggested terms and inter-relationship between them is shown in Table 7.2. The term chondrocalcinosis is a precise one that should be used to describe proven deposition of calcium salts within the cartilage. This may be seen on the radiographs, and the appearance can be described according to the site, density, and distribution of the shadow. Pyrophosphate deposits characteristically form linear or stippled lines of calcific density within fibrocartilage. If synovial fluid analysis reveals the presence of crystals, these should be described as outlined in Appendix I; pyrophosphate crystals usually exhibit weak positive birefringence. Any associated arthropathies can then be documented, and related to the radiographic and synovial fluid findings. This approach should avoid confusion.

7.3   The metabolism of inorganic pyrophosphate

Inorganic pyrophosphate is produced throughout the body by metabolic pathways involving phospho-ribonucleotides. Adenosine triphosphate (ATP), is degraded into adenosine diphosphate (ADP) and adenosine monophosphate (AMP) in basic energy-producing reactions within most cells and at cell surfaces. In addition, through the enzyme ATP pyrophosphohydrolase, cyclic AMP can be formed, particularly on cell

Table 7.2   The terminology of cartilage calcification

Cartilage calcification = chondrocalcinosis
In joints it may involve:
  fibrocartilage
  hyaline cartilage
Deposited salts may include:
  calcium pyrophosphate dihydrate
  hydroxyapatite
  dicalcium phosphate dihydrate
  mixtures
  ?others
It may be associated with:
  radiological evidence of calcification
  (often called chondrocalcinosis)
  acute arthritis
  (called pseudogout if pyrophosphate is deposited)
  chronic arthritis
  (called chronic pyrophosphate arthropathy, if
  pyrophosphate is deposited)

(The different types of arthritis are also called 'calcium pyrophosphate dihydrate crystal deposition disease' in the USA)

membranes, where it functions as a local messenger hormone, initiating other important metabolic pathways. As shown in Fig. 7.2, the formation of

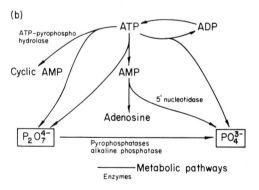

(a) $2H_3PO_4 \longrightarrow H_4P_2O_7 + H_2O$

Orthophosphoric   Pyrophosphoric   Water
acid   acid

(b)

ATP-pyrophospho hydrolase
ATP   ADP
Cyclic AMP   AMP
             5' nucleotidase
Adenosine
$P_2O_7^{4-}$ Pyrophosphatases alkaline phosphatase PO_4^{3-}

——— Metabolic pathways
Enzymes

Fig. 7.2   Metabolism of inorganic pyrophosphate: a simplified diagram of some of the pathways involved.   (a) Production from orthophosphate; (b) Metabolism of adenosine triphosphate (ATP).

ATP and of the base adenosine within these pathways liberates ortho-phosphate. However, breakdown of ATP to AMP or cyclic AMP results in the liberation of pyrophosphate which comprises two linked phosphate groups. This is normally further hydrolysed to orthophosphate by widespread pyrophosphatases including alkaline phosphatase.

The total daily turnover of pyrophosphate may be in the order of kilograms, and the plasma turnover time is about two minutes. Cell-free plasma levels of pyrophosphate are normally about 2 μM. However, normal levels varying by a factor of two have been reported by different authors, probably as a result of pyrophosphate release by platelets during venepuncture, or differences in the analytical methods used. Not only do problems arise in the accurate measurement of inorganic pyrophosphate, but great care has to be taken to use no tourniquets and or other factors that might damage platelets (a potent source of pyrophosphate) during collection of samples. Normal synovial fluid levels are similar to those in plasma, and the urinary concentration varies from 10 to 100 μM.

Pyrophosphate is not known to have any direct metabolic function, although it has been suggested that it may have a role in the inhibition of bone growth (Chapter 8). In common with other tissues the joints produce inorganic pyrophosphate. A number of authors have compared serum, urinary and synovial fluid levels in normal individuals and in patients with a variety of joint diseases. It is generally agreed that no joint disease is associated with an alteration in the plasma or urinary concentration of inorganic pyrophosphate, suggesting that any metabolic abnormality must be local rather than general. It is also widely accepted that patients with evidence of pyrophosphate deposition, as well as patients with osteoarth-ritis, have raised levels of inorganic pyrophosphate in the synovial fluid, the increase being in the order of a factor of 10. These raised synovial fluid levels could result either from increased production of pyrophosphate or from decreased hydrolysis. Several workers have looked, therefore, for evidence of abnormalities in inorganic pyrophosphatase activity in blood cells, as well as in the synovial fluid. However, no difference between control subjects and joint disease patients has been found, suggesting that the raised pyro-phosphate level is principally due to over-production. In patients with pyrophosphate chondrocalcinosis (Fig. 7.3), there are three possible origins of this raised level: (1) subchondral bone; (2) chondrocytes; and (3) syn-oviocytes. The work of Howell and others suggests that calcified cartilage produces about 10 pmol/mg dry weight/h of pyrophosphate (PP_i). The amount of pyrophosphate in bone is relatively large, and this may account for the increase in pathological tissues. It is also possible, however, that the cells within the joint tissues, and in particular the chondrocytes, are over-producers. Several groups have paid particular attention to the production of pyrophosphate from the cartilage. Lust *et al.* (1976) were unable to show

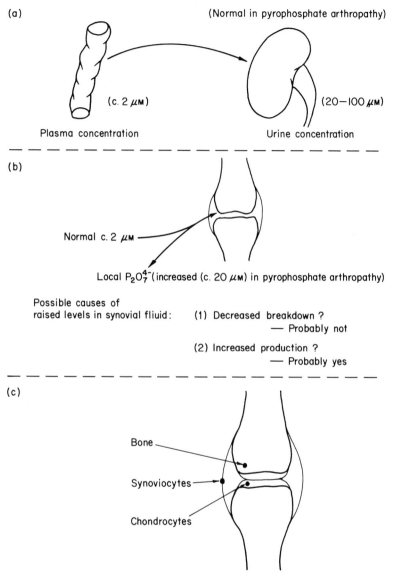

Fig. 7.3 Aspects of the metabolism of inorganic pyrophosphate ($P_2O_7^{4-}$). (a) Plasma levels and urine excretion of $P_2O_7^{4-}$. (b) $P_2O_7^{4-}$ in synovial joints. (c) Possible sources of joint $P_2O_7^{4-}$.

any difference in the extra-cellular production from cultured monolayers of chondrocytes taken from cartilage with and without pyrophosphate deposition, although the cells did accumulate PP_i. In contrast, Howell's group (1975), using micropipette techniques to obtain fluid from cultured segments of cartilage, have found evidence of a large increase in pyrophosphate production in the pathological specimens. Howell (1981) suggests that alteration in cells produced by separating out chondrocyte monolayers might explain the difference in the results. Furthermore, he has found evidence of increased activity of the enzyme 5'-nucleotidase, which breaks down adenosine monophosphate, suggesting that there is a general increase in the turnover. Seegmiller and colleagues (1981) have produced evidence for overproduction of PP_i by fibroblasts and leucocytes as well as joint tissues in some cases of pyrophosphate arthropathy. McGuire *et al.* (1980) have found that the chondrocytes can produce large quantities of intra-cellular pyrophosphate, but suggest that this cannot be extruded into the extra-cellular fluid unless the cells die. This group also showed that increased proteoglycan synthesis resulted in increased PP_i production by the cells.

Further study of these metabolic pathways may well lead to the elucidation of enzyme abnormalities and metabolic faults at a local level in pyrophosphate disease, analogous to those found in general purine metabolism in patients with gout. However, the localization makes it less feasible to treat the disease with a general drug analogous to allopurinol. The reader is referred to the recent review of McGuire *et al.* (1980) for further details of the available literature on the metabolism of PP_i.

7.4   The crystallization of calcium pyrophosphate dihydrate

The solubility and crystallization conditions for calcium pyrophosphate dihydrate are much more difficult to determine than those of the urates for a number of reasons. The pyrophosphate ion is sensitive to degradation, and is subject to rapid metabolic turnover so that reliable measurements of physiological concentrations are difficult to obtain. Pyrophosphate $(P_2O_7^{4-})$ is in equilibrium with the two acid forms $(HP_2O_7^{3-})$ and $(H_2P_2O_7^{2-})$ with pK values of 9.42 and 6.68. This means that at pH 7.4 the $(HP_2O_7^{3-})$ ion is the predominant form and the concentration of $(P_2O_7^{4-})$ is very sensitive to pH. Further, multiple-charged ions such as pyrophosphate and calcium are very prone to complex with species of the opposite charge thus reducing their effective activity in solution. Finally, calcium pyrophosphates occur in a variety of crystal forms. *In vivo* it appears that the monoclinic phase precipitates, and then slowly converts to the more stable triclinic form (Fig. 7.4). This is hard to duplicate *in vitro*.

A careful analysis of the available data on calcium formation of soluble calcium-pyrophosphate complexes and acidic pyrophosphates, yields a solubility product $(Ca^{2+})^2(P_2O_7^{4-}) = 3 \times 10^{-18}$ which can be regarded as accurate to about an order of magnitude (Brown and Gregory, 1976).

Physiological calcium levels are about 9 mg/100 ml (2.4 mM) of which in

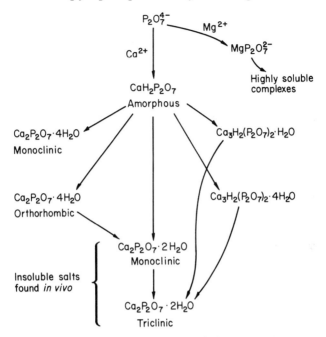

Fig. 7.4 Some calcium pyrophosphate salts and their possible inter-conversions (after Hearn and Russell, 1980).

synovial fluid 40% is estimated to be in the free, ionized, form compared to 60% in plasma. These free calcium levels of around 3.6 mg/100 ml (0.9 mM) to 5.4 mg/100 ml (1.4 mM) correspond to total pyrophosphate levels at saturation of 1–2.5 μM compared to measured levels in synovial fluid of 1–4 μM normally and 5–60 μM with pseudogout. Thus we conclude that normal fluids are in the region of saturation, whilst pseudogout fluids are supersaturated.

Early measurements of pyrophosphate crystal solubility in synovial fluid gave very high values but later studies produced levels in rough agreement with the synthetic studies. Magnesium, at physiological concentrations, forms stable complexes with pyrophosphate which greatly enhances the solubility. Such complexation might be expected to increase the solubility of pyrophosphate in synovial fluid and plasma. However, these fluids contain a range of charged macromolecules which are also capable of complexing with magnesium. This will be particularly true of the proteoglycans of the cartilage and hyaluronate. In fact there is almost no multiple-charged species which will not interact to some extent with calcium and magnesium or with pyrophosphate, as well as with other species of the opposite charge. Thus the synthetic solutions give a reliable measure of the true solubility in terms of free calcium and pyrophosphate, but to know the total solubility in synovial fluid we have to make direct measurements on synovial fluid itself.

The values that are available suggest there is not a lot of extra complex formation with pyrophosphate, but more data are needed.

Recently, Hearn and Russell (1980) have succeeded in growing crystals of orthorhombic calcium pyrophosphate tetrahydrate from solutions containing 1.5 mM calcium and 40 μM pyrophosphate which are about 40 times the saturation level of pyrophosphate but within the range reached in pseudogout. Magnesium hinders this precipitation but phosphate counteracts the magnesium.

Thus normal synovial fluid seems to be about saturated for pyrophosphate whilst in pseudogout levels vary through the metastable region to the region of the supersolubility limit (see Fig. 3.6). Exact pyrophosphate levels in articular cartilage are not known, and probably vary at different depths. The observation of crystal deposition in this tissue suggests that factors aiding nucleation may be present within the cartilage. Although there is, as yet, no evidence that cartilage components aid this nucleation, it has been shown by Russell's group that iron does increase crystal formation *in vitro*. Thus a change in pH, an increase in iron levels in joints or other local cartilage changes may predispose to the formation of precipitates.

7.5 The crystals

The crystals that deposit *in vivo* have been identified by powder X-ray diffraction as the monoclinic and triclinic forms of calcium pyrophosphate dihydrate; the triclinic salt usually predominates. In the polarized light microscope the crystals appear as small rods or prisms, varying in length from 1 to 10 μM, and in width from 0.2 to 5 μM. They usually exhibit weak positive birefringence, and often appear to have a chip out of one corner, due to twinning of two unequally sized crystals (Fig. 7.5 and colour Plate 5).

Examination in the electron microscope shows up more interesting features of these crystals. They vary considerably in size and shape, the smallest being well under a micron in length, and numerous growth lines can be observed. Twinning and the formation of a variety of odd morphological shapes is not unusual, as shown in the illustrations (Fig. 7.6). Like urate crystals, pyrophosphate crystals tend to become foamy on examination in the electron microscope, presumably due to the loss of water of crystallization. Another interesting finding has been the presence of large numbers of fractured crystals in calcified cartilage. This suggests that the stresses on the cartilage can produce new surfaces for nucleation of further crystals, and increase the damage done. Extensive studies of the nature of the growth lines and internal structure, as has been carried out by Pritzker for urate crystals, are at present under way and may give further clues to the formation and the nucleation and growth of these salts. Scanning electron micrography has revealed evidence of surface attachment of small deposits of a granular

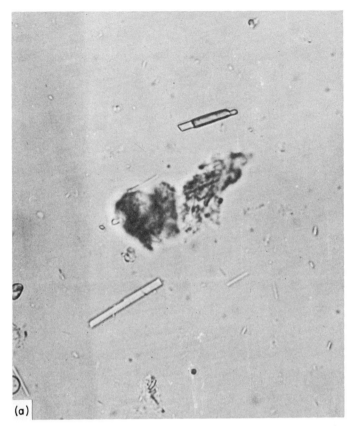

Fig. 7.5   (a) Calcium pyrophosphate dihydrate crystals in synovial fluid, viewed in
the polarized light microscope. Note the morphology, and twinning of
crystals, which often makes them appear to have a chip out of one corner
($\times 600$). (b) See Colour Plate 5.

material, similar to those found on urate crystals, and these, presumably,
represent aggregation of protein to the surface occurring in some
circumstances *in vivo*. Studies with radiolabelled crystals suggest that they
have a half-life of about 3 months *in vivo*.

7.6   Conditions associated with the deposition of calcium pyrophosphate dihydrate crystals

7.6.1   *Metabolic diseases*

Further clues as to the nature of the local metabolic abnormalities in
pyrophosphate deposition come from systematic studies of associated

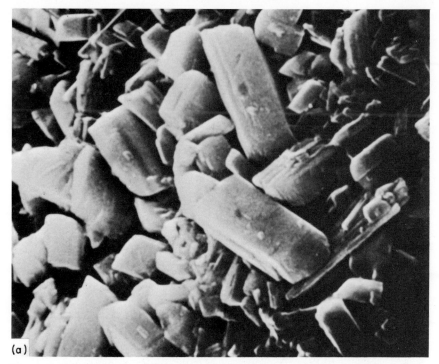

Fig. 7.6 (a) and (b) Scanning electron micrographs of calcium pyrophosphate dihydrate crystals from joint cartilage. Note the range of morphology and size of the crystals ((a) × 12 000, (b) × 2000).

conditions found in people with deposits.

A bewilderingly large number of both generalized and local abnormalities have been described in patients with pyrophosphate deposition (Table 7.3). The problem in attaching significance to these associations arises because pyrophosphate deposition is such a common occurrence, as are several of the suggested associations. Controlled studies are required, and there are very

Table 7.3 Some of the conditions associated with deposition of pyrophosphate crystals

General	Local
Hyperthyroidism	Joint instability
Haemachromatosis	Previous meniscectomy
Hypothyroidism	Osteochondromatosis
Gout	Amyloidosis
Hypomagnesaemia	Biochemical changes in
Hypophosphatasia	the matrix
Steroid therapy	

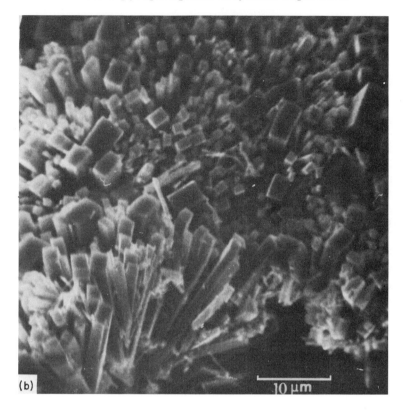

(b)   10 μm

few in the literature. McCarty and his colleagues (1974) reported a controlled study of disease associations in 28 patients with crystal deposition, and Ellman *et al.* (1980) have recently reported a study involving nine biochemical tests carried out on 36 patients and compared with 29 closely matched controls. These small but carefully controlled studies reveal few differences in the patients, and only hyperparathyroidism emerges as an occasional metabolic association.

We (Alexander *et al.*, 1982) have recently reviewed 105 patients with pyrophosphate arthropathy, and compared their associated metabolic conditions and drug therapy with those of 105 age–sex matched admissions to a general medical ward. Metabolic screening tests were carried out on the pyrophosphate patients and on 48 consecutive cases of osteoarthritis presenting at the same clinic. Positive associations of pyrophosphate arthropathy to emerge included hypothyroidism, hyperparathyroidism and steroid therapy (Table 7.4).

The other approach to define metabolic associations is to study patients with a known disease, and compare them with a control group for the

Table 7.4   Disease associations in 105 consecutive patients with pyrophosphate and arthropathy, and comparison groups of 105 age–sex matched medical patients and 48 patients with osteoarthritis. Results expressed as a percentage

	Diagnosis (number)		
	Pyrophosphate arthropathy (105)	Osteoarthritis (48)	Acute medical admission (105)
Rheumatoid arthritis	7.8	0	1.5
Hypothyroidism	10.5	2.1	2.8
Hyperthyroidism	3.8	2.1	2.8
Hyperparathyroidism	1.9	0	0
Chronic steroid therapy	15.2	0	2.4
Gout	1.9	0	2.4
Hypertension	18.1	8.3	15.2
Diabetes mellitus	2.8	0	7.6

Reproduced with permission from Alexander *et al.* (1982).

incidence of pyrophosphate deposition (usually defined radiologically). To date, this approach seems to have been more valuable, and in many cases solid associations have emerged. These include:

(a) *Hyperparathyroidism.* There is a marked increase in the incidence of chondrocalcinosis in patients with hyperparathyroidism (Pritchard and Jessop, 1977). Furthermore, they tend to develop crystal formation much younger than patients without any obvious metabolic abnormality. Crystal deposition appears to be proportional to the duration of disease, and its severity. Strangely, operations to remove the parathyroids not only induce attacks of pseudogout, but do not appear to remove the tendency to crystal deposition, which increases rather than decreases. Hyperparathyroidism causes hypercalcaemia, through the multiple actions of parathormone; the presumed fault is an increase in local calcium concentration in the articular cartilage, but parathormone may also increase pyrophosphate production by cells. The relative lack of an association between chondrocalcinosis and other hypercalcaemic states, supports evidence for an important role for parathormone itself. The lack of 'cure' by parathyroidectomy raises doubts, but it may be that both prolonged hypercalcaemia and raised levels of parathormone are required to induce crystal formation, and that once a large deposit is present, further crystal aggregates easily form, so that progress cannot then be stopped by removing the cause.

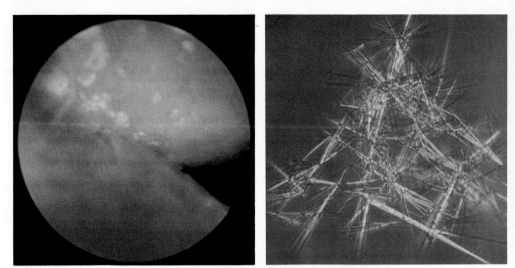

Plate 1

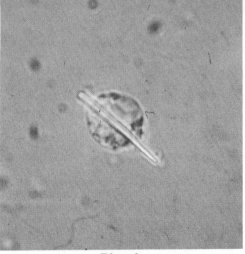

Plate 3

Plate 2

Plate 1 (Fig. 4.4) Arthroscopic view of synovium in pyrophosphate arthropathy showing white crystal deposits in the inflamed synovial fronds (× 2.5).

Plate 2 (Fig. 4.11) Urate crystals viewed through the polarized light microscope with the first-order red compensator (× 400).

Plate 3 (Fig. 6.10) A urate crystal in synovial fluid. Note the needle-shaped morphology, negative birefringence, and attachment to a phagocytic cell (polarized light microscopy, high-power field) (× 600).

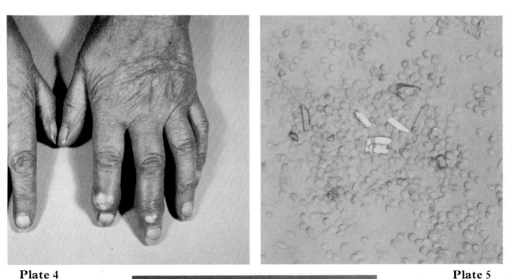

Plate 4   Plate 5

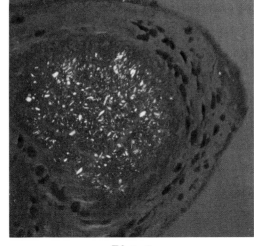

Plate 6

Plate 4 (Fig. 6.14(c)) Chronic tophaceous gout of the hand showing white deposits of monosodium urate monohydrate crystals in the skin and subcutaneous tissues overlying the distal interphalangeal joints.

Plate 5 (Fig. 7.5(b)) Pyrophosphate crystals in synovial fluid, viewed in a polarized light microscope. When a first order red compensator is used the crystals are seen to show positive birefringence. In this blood-stained fluid, the crystals are surrounded by erythrocytes ($\times 600$).

Plate 6 (Fig. 7.8) Histological section of synovium showing a deposit of pyrophosphate crystals surrounded by inflammatory cells ($\times 40$).

A number of authors have suggested that raised levels of parathormone exist in some patients with pyrophosphate deposition in the absence of obvious hyperparathyroidism; this is now disputed, and difficulties in accurately measuring levels of parathormone make some of these studies open to criticism.

(b) *Haemachromatosis.* Haemachromatosis is a relatively rare disease associated with a marked excess of iron stored in connective tissue. There is a clear, marked increase in the incidence of chondrocalcinosis in these patients. A large group has been carefully studied by Hamilton and his colleagues (Atkins *et al.*, 1970). Extensive pyrophosphate deposition occurs, and may be associated with a distinctive arthropathy. As in the case of hyperparathyroidism, removing the stimulus, i.e. treating the patient adequately with venesection, does not apparently prevent the progression to further formation of pyrophosphate deposits. The mechanism in this case may be via excess local iron acting as a nucleating factor for pyrophosphate crystal deposition, or acting as a poison of pyrophosphatase activity; it is of interest that we have recently observed patients in whom an infusion of iron induced florid pseudogout.

(c) *Hypophosphatasia.* A few cases of the very rare condition hypophosphatasia have been reported in association with premature deposition of pyrophosphate and recurrent attacks of arthritis. This condition is characterized by very low levels of alkaline phosphatase, and by increased excretion of phenyl ethanolamine in the urine. The presumed defect leading to pyrophosphate deposition is inadequate hydrolysis of inorganic pyrophosphate to inorganic phosphate, due to lack of local alkaline phosphatase.

(d) *Hypomagnesaemia.* There have been several reports of premature pyrophosphate deposits occurring in patients with excessively low magnesium levels. The cases reported include idiopathic hypomagnesaemia, as well as other conditions associated with low magnesium salts, such as Baarter's syndrome. Furthermore, many of the stimuli which are associated with the production of attacks of pseudogout are also associated with transient low magnesium levels; for example following surgery. As mentioned, magnesium will complex with inorganic pyrophosphate to form highly soluble complexes. A relative lack of magnesium at the local level may predispose to the formation of the calcium salt in preference, and this of course is much less soluble.

(e) *Gout.* There have been a number of associations of gout, and pyrophosphate deposition, and in a recent study Scott and his colleagues

(Stockman *et al.*, 1980) produced evidence of an increased incidence of pyrophosphate deposition in gouty subjects, compared with a control population. The cause of this association is not obvious. Hydroxyapatite is often found in gouty tophi, and the formation of pyrophosphate and hydroxyapatite may be due to the urate crystals acting as epitaxial nucleation sites for the formation of calcium phosphates. Alternatively, it may be due to an increase in phosphate turnover in the region damaged by urate crystals, or another secondary effect of the urate.

(f) *Hypothyroidism.* Several authors have described chondrocalcinosis in association with thyroid deficiency, and in our survey this was one of the most frequent abnormal findings (Table 7.4). The reasons for this are not yet clear, although hypothyroidism is known to cause changes in connective tissue, which may predispose to local crystallization.

(g) *Steroid therapy.* In our series of 105 patients with pyrophosphate arthropathy there were 17 who had been on long-term steroid therapy (> 7.5 mg prednisolone for > 5 years). Several patients had been put on prednisolone for joint symptoms, but in many the therapy apparently preceded the appearance of chondrocalcinosis, suggesting that this may be another metabolic predisposition, although this needs to be substantiated by other studies and could be a selection artefact.

Other metabolic associations may include ochronosis and Wilson's disease.

All surveys show that only a minority of patients have any of these metabolic predispositions. Patients found to have chondrocalcinosis should probably be screened for thyroid disease and hyperparathyroidism, but more thorough metabolic screening, in the absence of clinical abnormalities, is unlikely to be rewarding.

7.6.2　Local abnormalities

Chemical factors predispose to the formation of pyrophosphate in a minority of cases; in others a mechanical or physical abnormality apparently precedes the deposition of crystals at a local level.

Bird and colleagues (1978) first described the association of chondrocalcinosis and joint laxity (hypermobility), and in our series this was a common finding (11.5%). There are two possible explanations: (1) joint laxity or instability might cause local damage, predisposing to deposition; (2) hypermobility may indicate a generalized abnormality of connective tissue which also favours pyrophosphate formation. We have observed several patients in whom the deposits were isolated to unstable joints, favouring the first hypothesis. One of the most striking is a 34 year-old lady with one unstable knee; she developed chondrocalcinosis in this knee only, in the absence of any other apparent predisposing factor.

Other local abnormalities may cause chondrocalcinosis: patients with a past history of a meniscectomy have developed chondrocalcinosis in the operated knee only, and we have seen cases of peroneal muscular atrophy with pyrophosphate deposition limited to their weak, unstable ankle joints.

Many authors have also reported a high incidence of generalized osteoarthritis in pyrophosphate arthropathy, and an association with rheumatoid arthritis. It is often difficult to know which condition comes first, to know if one is causally related to the other, or whether these are chance associations. Our own series suggest a high incidence of local joint abnormalities of one sort or another preceding the appearance of chondrocalcinosis.

Recent pathological studies have suggested that there may be alterations in the chemistry of the cartilage matrix in patients with primary pyrophosphate arthropathy (Bjelle, 1981). Furthermore, local changes in the cells and matrix of cartilage and synovium may explain the recently described associations of amyloid deposits and osteochondromatosis with pyrophosphate deposition (Wilson and Irvin, 1981). The associations with gout, hypothyroidism and steroid therapy could be explained by similar changes.

It seems reasonable to suppose that further elucidation of local factors in the cartilage matrix which might determine the initiation and growth of pyrophosphate crystals are as likely to lead to further understanding of this disease as is the search for a general metabolic defect.

7.6.3   The epidemiology of pyrophosphate deposition and its association with age

Estimates of the prevalence of pyrophosphate deposition vary considerably. Data have been collected in two ways: (1) by radiological surveys, which depend on the type of film used, the radiographic techniques, and on subjective interpretation of the shadows seen; and (2) by the pathological examination of cadaveric material, which will depend on the extraction of visible deposits and their analysis by crystallographic techniques. A summary of some of the data obtained in this way is shown in Table 7.5. The two most reliable anatomical studies gave prevalences of 3.2% and 6.8% of pyrophosphate deposition in meniscal cartilage. The radiological studies give far more variable results, although a good deal of this can be explained by differences in technique, or by differences in the age of the subjects. Most surveys give a radiological prevalence of around 5%, increasing in the elderly. A survey of Ellman and Levin (1975) stands out from this, recording an incidence of 27.6% in elderly Jewish subjects living in an old people's

Table 7.5 Prevalence of pyrophosphate deposits in the knee

	Reference	No. studied	CPPD found
Pathological	McCarty et al. (1966)	215	3.2%
	Lagier and Baird (1968)	320	6.8%
Radiological	Wolke (1935)	?	0.19%
	Bocher et al. (1965)	455	7.0%
	Cabanal et al. (1969)	200	6.5%
	Zinn et al. (1969)	131	4.6%
	Pritchard and Jessop (1977)	100	11%
	Ellman and Levin (1975)	58 (elderly)	27.6%
	Wilkins et al (1982)	100 elderly	34%

home. We (Wilkins *et al.*, 1982) have recently surveyed 100 consecutive admissions to a local geriatric hospital and found a similar high incidence of radiological evidence of pyrophosphate deposition (34%). There was a clear association with increasing age, the prevalence rising to 47% in patients over the age of 85.

Bywaters (1979) has recently studied the incidence of calcification in interspinus bursae, the ligamentum flavum and discs. In an elderly population sample, pyrophosphate was found in 14% of spines studied, mainly within fissures in the discs. Hydroxyapatite was found in 20%. In neither case was there any obvious relationship between the presence of the deposits and inflammatory or degenerative changes. An association with age was again apparent, and it is clear that age alone can often be associated with deposition in the absence of any obvious associated joint disease.

7.6.4 Familial forms of pyrophosphate deposition

Several large series of familial pyrophosphate arthropathy have been reported from a variety of countries including Czechoslovakia, Chile, Holland and France. Most series have shown male-to-male transmission, strong penetrance, and severe associated arthritis, with appearance of chondrocalcinosis early in life, and no other factor predisposing to deposition. This suggests that an autosomal dominant inheritance may operate. In the series of sporadic cases described by McCarty, by ourselves and others, positive family histories have not been common, and familial cases have not been seen. However, in the absence of a careful examination of relatives no definite conclusions can be reached about the incidence of a genetic predisposition to chondrocalcinosis.

Table 7.6 Classification of pyrophosphate deposition

(1) Familial forms:
     Czechoslovakian
     Chilean
     Dutch
     Others
(2) Sporadic (no obvious cause)
(3) Associated with metabolic disease (e.g. hyperparathyroidism)
(4) Associated with local joint problem (e.g. hypermobility or amyloidosis)
(5) Age-associated

7.7 Classification of pyrophosphate arthropathy

Many authors have classified the disease into familial, metabolic, age-associated and sporadic forms. We believe that cases associated with local joint damage predisposing to deposits form a further separate group (Table 7.6).

The advantage of such a classification is that it highlights the differences between these cases. Familial cases often develop very premature chondrocalcinosis, followed by the progressive development of the arthropathy; some metabolic diseases behave similarly. In many of the sporadic forms, or those associated with a minor metabolic or local abnormality, the disease starts later, but a progressive destructive arthropathy may then develop. In contrast, age-associated chondrocalcinosis is relatively free of severe joint damage. The survey of geriatric patients showed that many of these with chondrocalcinosis had no arthritis, although there was a statistically significant correlation between local joint damage and pyrophosphate deposits.

Men and women with chondrocalcinosis also show differences in the expression of the disease. Men tend to be younger, to have the condition isolated to the knees, to get more acute arthritis, and less destructive changes. The women have a wider distribution of disease, and more chronic arthritis.

7.8 The pathology of pyrophosphate deposition

Pyrophosphate deposits may be found in hyaline joint cartilage, in the synovial fluid and synovium of joints, and in fibro-cartilage pads such as the symphisis pubis and inter-vertebral discs. It is occasionally found in periarticular sites such as tendon insertions, but there are very few cases in literature of pyrophosphate deposition away from the joints or their immediate vicinity; in one case crystallographic analysis revealed the presence of calcium pyrophosphate dihydrate in the meninges of the brain (Grahame and Sutor, 1971).

The deposits seem to have a peculiar preference for fibro-cartilage over hyaline cartilage. The commonest sites are the knee menisci, the triangular ligament of the wrist, the symphisis pubis and the inter-vertebral discs, all of which consist primarily of fibro-cartilage. Hyaline articular cartilage is the next most commonly involved structure, and the knee is again the commonest joint, although the hips, elbows, wrists and small joints of the hands and feet may also be involved. Arthroscopic studies reveal evidence of synovial deposits, and rare cases have been described in which synovial deposition is apparent in the absence of any obvious cartilage deposits. As already mentioned, most workers believe that the cartilage is the primary site of crystallization, and initial precipitation within the synovial fluid or elsewhere seems unlikely, although deposits may be 'seeded' there.

Within the cartilage itself, deposits can be seen in three sites. (1) In the perichondrocyte region (2) the mid zones and (3) superficially. The largest deposits, and the most common site of deposition is the centre of fibro-cartilage pads or in the mid-zone of hyaline articular cartilage (Fig. 7.7). The crystals often form large sheets of deposits within a granular matrix, resulting in the familiar linear shadow on X-ray. Discreet rounded accumulations can also be seen, and are reflected by the stippled pattern that can also be detected radiographically. The perichondrocyte deposits are less common, and usually consist of very small crystals (some 250–500 Å long). They are not seen in the matrix vesicles, which is perhaps not surprising if high concentrations of alkaline phosphatase are found there, but electron microscopy has recently revealed the presence of crystals in close association with chondrocytes, strengthening the case for crystallization at the site of over-production of pyrophosphate from within the chondrocytes. On this basis, the superficial deposits that are also seen may grow from crystals coming initially from the deeper sites, but this is not proven.

A large variety of pathological changes may be seen in the cartilage, synovium and periarticular tissues in association with these deposits. Very often there is no obvious change. In other instances, destructive, degenerative changes in the cartilage occur, and in others there is a very florid, severe destructive osteoarthritis that may progress to a 'pseudo-Charcot joint'. Synovial changes are also variable: deposits often occur in relatively avascular areas with no apparent inflammation; but in other cases, a giant cell or low grade inflammatory reaction may be seen around the crystals, and some show a widespread inflammatory reaction dominated by polymorphonuclear cells. There appears to be no constant relationship between the presence of deposits, inflammation, or osteoarthritic changes (Fig. 7.8–see colour Plate 6).

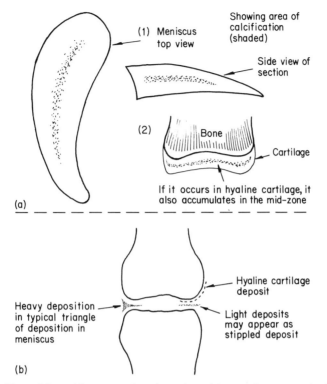

Fig. 7.7   Sites of deposition: pyrophosphate deposition tends to occur in the mid-zone of meniscal fibro-cartilage, but can also occur in articular hyaline cartilage. (a) Site of deposit, (1) fibro-cartilage menisci, (2) hyaline cartilage. (b) X-ray appearances.

7.9   Associated clinical conditions

The number of clinical syndromes that have been associated with the presence of pyrophosphate crystals is bewilderingly large, almost every type of arthritic disease has been described. The problem is to distinguish the real associations from the chance ones; pyrophosphate deposition having a high prevalence, particularly in an elderly population.

McCarty and his colleagues (1976, 1977) have split the clinical syndromes into five: (1) acute arthritis (pseudogout); (2) osteoarthritis; (3) osteoarthritic changes with intermittent acute attacks; (4) pseudo-rheumatoid arthritis; and (5) pseudo-neuropathic joints. Although this typing is widely followed, it may obscure the true relationship between disease and pyrophosphate deposits. In our own surveys of consecutive patients observed to have arthritis and pyrophosphate deposition, in addition to those found to have pyrophosphate deposition on a radiological survey, we have identified the two basic patterns of acute inflammatory arthritis and chronic destructive arthritis originally described. The severity and form of these has varied, and there has been considerable overlap with other conditions. Our patients have not otherwise conformed to the classification described (Dieppe *et al.*, 1982). For the sake of clarity and simplicity, we will describe the clinical associations under the two major headings, as is customary in the case of gout (acute gout or chronic tophaceous gout), and hydroxyapatite deposition (calcific periarthritis and synovitis, or osteoarthritis).

7.9.1   *Acute arthritis (Pseudogout)*

'Pseudogout' is a common and well-known accompaniment of pyrophosphate deposition. The attacks differ in many ways from gout: (1) they tend to be slower in onset, to take longer to resolve, and to be less severe, and (2) the sites of involvement are quite different, the 'knee being to pseudogout what the toe is to gout' (McCarty).

Acute pseudogout can be defined as an acute, self-limiting, attack of arthritis lasting anything from one day to four weeks, associated with a joint effusion containing crystals of calcium pyrophosphate dihydrate, which can usually be seen intra-cellularly in the polymorphonuclear leucocytes. The phenomenology of these attacks, based mainly on our own series is shown in Table 7.7 and is similar to that described by other authors. The knee is by far the commonest joint involved, and the average attack will last for about 10 days. It is self-limiting, and may respond to colchicine, although the therapeutic response is less predictable and less dramatic than in gout. There is an important difference between men and women with pyrophosphate deposition in the incidence type and distribution of acute attacks. As previously mentioned, males with this disease tend to be younger. They also tend to have more attacks of pseudogout involving more joints, although the knee is by far the commonest. The women are more prone to upper limb attacks and polyarticular involvement if they do get pseudogout. A precipitating event was apparent in about 20% of our patients, and, as found by other authors, included an inter-current disease or minor trauma. (These figures may, of course, be spurious because of the high pick-up rate of acute attacks in hospitals; many of these attacks may not have been diagnosed if they had occurred in a different setting with the patient at home).

Variations in acute attacks have led to descriptions of a number of other syndromes associated with pyrophosphate deposition (e.g. 'pseudo-rheumatoid arthritis', haemarthrosis, etc.). The important variance found in our own series is shown in Table 7.7. About a third of patients have more than one joint involved. The attack may start with one joint (usually the

Table 7.7   Acute arthritis in association with pyrophosphate deposition ('pseudogout').
Data from an unselected series of 105
consecutive patients with pyrophosphate arthropathy

(1) Number of patients suffering 'pseudogout'
attacks:

women:	28%
men:	54%

(2) Joints affected:

	men	women
knee	93%	71%
ankle	25%	4.8%
wrist/hand	13%	62%

(3) Mean duration:
approximately 10 days

(4) Recurrent attacks:

men	73%
women	40%

(5) Polyarticular attacks:

men	13%
women	24%

Variants of acute pseudogout

(1) Blood-stained synovial fluid
(2) Intercurrent infection
(3) Urate crystals also present
(4) Periarticular site of acute inflammation (e.g.
quadriceps tendon)
(5) Other syndromes described include:
'polymalgia rheumatica'
acute back pain

knee), and then spread to involve other sites. This is the so-called cluster attack. More rarely, several joints are involved all at once, or the attack flits round a number of sites, and this may lead to diagnostic confusion. If a fever, or raised erythrocyte sedimentation rate (E.S.R.) is present, rheumatoid arthritis or septic arthritis is occasionally diagnosed. As already mentioned, pyrophosphate deposition can occur in the presence of true rheumatoid arthritis, leading to further diagnostic confusion. In pseudogout, the synovial fluid is commonly blood-stained. This may be due to the type of inflammatory reaction, but seems more likely to be associated with its initiation by trauma, leading to release of the crystals (see below). Other diagnostic pitfalls include the presence of concurrent infection or gout, as well as pseudogout. We have seen four patients with an acute inflamed joint, in whom the synovial fluid was found to contain numerous pyrophosphate crystals, and a diagnosis of pure pseudogout was made. Fortunately, fluid was also sent to the bacteriology department, where Gram stain and culture showed the presence of an infection as well. Without the bacteriology, the more important diagnosis would have been missed. Periarticular deposits of pyrophosphate can also cause acute inflammation. As explained, the commonest sites are the tendon insertions, particularly at the quadriceps tendon and achilles tendon, and acute inflammation at both sites has been described. Periarticular inflammation in association with pyrophosphate crystals occasionally occurs elsewhere, such as the elbow, shoulder and wrist. X-rays will almost invariably reveal intra-articular linear calcification, as well as whiskers of calcification at the tendon insertion involved in the acute attack. Finally, pyrophosphate deposition has been associated with a polymyalgic-like syndrome, and acute attacks of back pain.

A high degree of clinical suspicion is required if this condition is to be diagnosed correctly ('if in doubt, think of pseudogout as well as gout!'). However, the majority of attacks occur in the knees of elderly people in association with radiological evidence of chondrocalcinosis and with pyrophosphate crystals visible in the synovial fluid. This is common and easy to diagnose, although often wrongly considered to be a flare-up of osteoarthritis or a traumatic effusion in the joint. In view of the prompt therapeutic response to anti-inflammatory drugs or aspiration of fluid and injection of steroid (if infection has been ruled out), an incorrect diagnosis can result in unnecessary pain, stiffness and damage to the joint.

7.9.2   The pathogenesis of pseudogout

The aetiology and pathogenesis of the acute attack are not entirely clear. Either acute crystallization or crystal shedding from a preformed deposit could initiate the inflammation. Crystal shedding seems more likely; the evidence in favour of this includes the initiation of acute attacks on washing out a joint with magnesium to solubilize the crystals, thus displacing them. In two of our own cases, disappearance of radiological calcification has accompanied the acute attack, strongly suggesting shedding of crystals (Doherty and Dieppe, 1981).

The crystals themselves probably produce acute inflammation in a similar way to urate crystals, as described in the previous chapters. They can activate complement, lyse red cells and are phagocytosed with consequent enzyme release. In each of these cases their effect is quantitatively less than that of urate crystals, and they also have a lower surface charge, and less tendency to pick up surface protein. The response would therefore appear to be qualitatively similar but quantitatively less, and this is mirrored in the less dramatic nature of the pseudogout attack compared with that of gout. Many patients have abundant pyrophosphate crystals in their synovial fluid, in the absence of any acute arthritis. This is much more striking than in gouty patients, and the cause for this remains unknown. Whether it is due to a limited proportion of patients having crystals which will pick up appropriate protein to stimulate the attack, or whether it is related to the size, nature and number of the crystals present remains unknown. Our own data would suggest that crystals tend to be smaller in the acute attack than in chronic pyrophosphate arthropathy, although this could also be due to the damage of preformed deposit or acute crystallization initiating the attack, rather than to the necessity for a small crystal to cause the inflammation. Further work on the ability of pyrophosphate crystals to induce acute inflammation may help in the understanding and treatment of the condition.

7.9.3   Chronic destructive arthritis

This is often clinically indistinguishable from generalized osteoarthritis ('pseudo-osteoarthritis') and is the other main clinical feature of pyrophosphate arthropathy. The characteristic features of this condition are shown in Table 7.8. In contrast to acute pseudogout, women are affected more than men. (This is partly due to the greater age of women in most series, as the chronic arthritis shows a clear correlation with increasing age.) The chief joints involved are the knees, wrists and hands, although shoulders and ankles are also commonly affected. The clinical features are basically those of osteoarthritis and include pain on use of the joint, crepitus and bony swelling, with, or without, a modest effusion, tenderness over the joint line, and limitation of movement.

A number of interesting variants of the chronic 'osteoarthritis' pattern have been observed. In many, the joint destruction is very severe, with radiological evidence of bone damage and large cyst formation, often in the absence of prominent osteophytes (Richards and Hamilton, 1974). Other variants are detailed in Table 7.8. In the knee, the destruction classically affects the medial side of the joint, with collapse of the tibial plateau being relatively common. The more rare syndrome of medial femoral necrosis has also been seen in association with pyrophosphate deposits. High tibial fractures, perhaps related to the stress on the medial joint line and the instability can also occur. Patello-femoral disease is also frequently severe and extensive. Destruction and subluxation of the wrist joint is not uncommon, and the combination of severe knee and wrist disease in an elderly lady is very suggestive of pyrophosphate arthropathy. Shoulders, elbows and ankles may be involved, and bone destruction may be so severe as to mimic a Charcot's joint ('pseudo-neurotrophic arthritis'). Occasionally

Table 7.8 Chronic arthritis in association with pyrophosphate deposition (*'chronic pyrophosphate arthropathy'*). Data from an unselected series of 105 consecutive patients

---

(1) Number of patients with chronic symptomatic arthritis:
  men   (63%)
  women (97.3%)
(2) Incidence increases with increasing age.
(3) Joints involved in order of frequency

	(%)
Knee	92
Distal-interphalangeal	73
Carpo-metacarpal	53
Wrist	23
Shoulder	39
Ankle	32

(4) Number with a pattern indistinguishable from 'generalized osteoarthritis'
  men   17.8%
  women 53.3%
(5) Number with severe destructive joint disease
  men   2%
  women 28%

Other features of chronic arthritis in association with pyrophosphate deposition

---

(1) Osteoarthritis predominantly in the knee and wrist, with severe patello-femoral disease
(2) Severe destructive changes in large joints with large bone cysts and few osteophytes on the radiograph
(3) Collapse of medial tibial plateau *or* medial femoral necrosis of the knee joint
(4) High tibial fractures in association with severe angulation deformity of the knee
(5) Severe patello-femoral disease with florid new bone formation
(6) 'Pseudo-neurotrophic joints'
(7) Spinal collapse and destruction

---

spinal disease may progress to destructive changes with loss of height and severe back pain.

An analysis of the distribution and type of changes seen in the hands revealed a high incidence of 'nodal' swellings of the distal and proximal inter-phalangeal joints, plus carpo-metacarpal joint damage. This is the pattern typical of generalized osteoarthritis (OA). The nature of the association between these two conditions is not clear. The three main possibilities are: (1) a chance association; (2) OA precedes and predisposes to pyrophosphate deposition; (3) pyrophosphate deposits precede and cause OA. In elderly patients an association was found between the presence of chondrocalcinosis and OA; similarly the cases of severe destructive disease in the presence of crystals argue that they can 'cause' a chronic destructive arthropathy. However, in many cases OA seems to precede chondrocalcinosis, and crystals can be deposited for years without there being any significant joint damage.

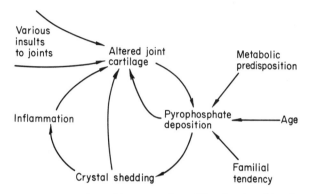

Fig. 7.9   Pyrophosphate deposition as an 'amplification loop' in chronic arthritis.

One hypothesis resolving this argument is shown in Fig. 7.9. Local damage (such as minor degrees of OA) may predispose to pyrophosphate deposition. The presence of crystals may then accelerate damage. Thus crystal deposits would act as an 'amplification loop' in chronic arthritis. The hypothesis may also explain the relationship between Charcot joints and chondrocalcinosis. The joint destruction in pyrophosphate arthropathy is often severe enough to mimic neurotrophic joints, and in one small series pyrophosphate was identified in patients with true syphilitic Charcot joints (Bennett *et al.*, 1974). If deposits may form in damaged joints, and accelerate destruction, these findings could then be explained.

7.9.4   The pathogenesis of destructive changes

Our hypothesis supposes that pyrophosphate crystals are especially damaging to an already abnormal joint.

The mechanisms that might operate have been discussed in Chapter 5. The effect of calcific deposits on cartilage compliance, and the ability of the crystals to cause surface wear, may be particularly important. Recent work showing that synoviocytes in culture may release collagenase and other proteases in the absence of acute inflammation may also be relevant. However, very little work has been done on this subject, in spite of the relatively frequent, and very severe, joint damage seen in association with chondrocalcinosis.

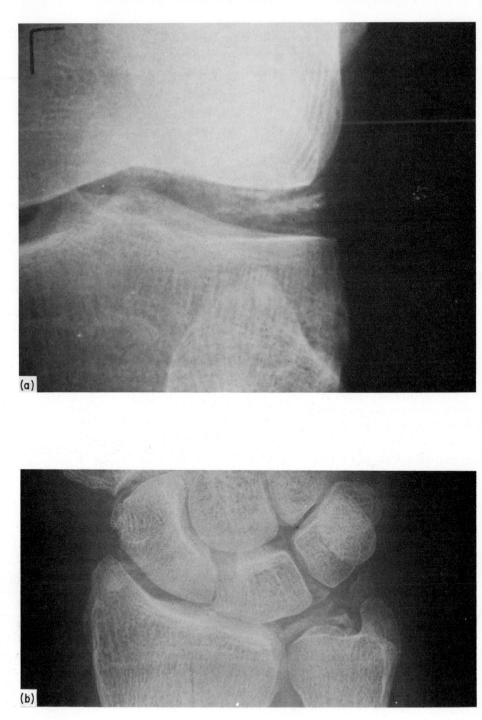

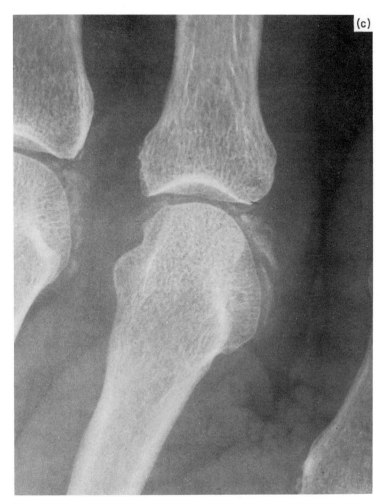

(c)

Fig. 7.10 Radiographs of chondrocalcinosis (a) Chondrocalcinosis of the lateral compartment of the knee joint showing the typical linear calcification in the menisci, and hyaline cartilage. (b) Chondrocalcinosis of the triangular ligament of the wrist joint. (c) Chondrocalcinosis in a metacarpophalangeal joint.

181

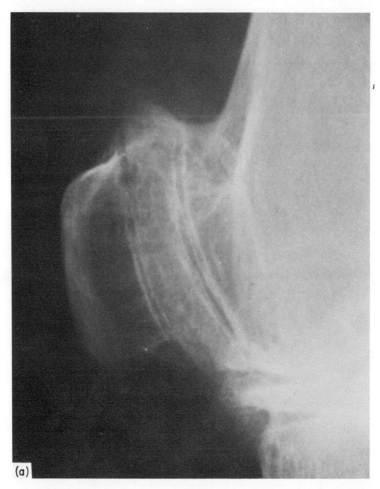

(a)

Fig. 7.11 Radiographs of chronic pyrophosphate arthropathy (a) Lateral radiograph of the knee joint showing extensive new bone growth at the patello-
femoral joint; a classical feature of pyrophosphate arthropathy. (b) Knee
joint in chronic pyrophosphate arthropathy showing gross destructive
changes in the lateral tibial plateau, and obliteration of the joint space
obscuring the chondrocalcinosis. (c) Pyrophosphate arthropathy of the
wrist. Note the gross cystic changes, bone destruction at the end of the
ulna, and sclerosis without osteophytes.

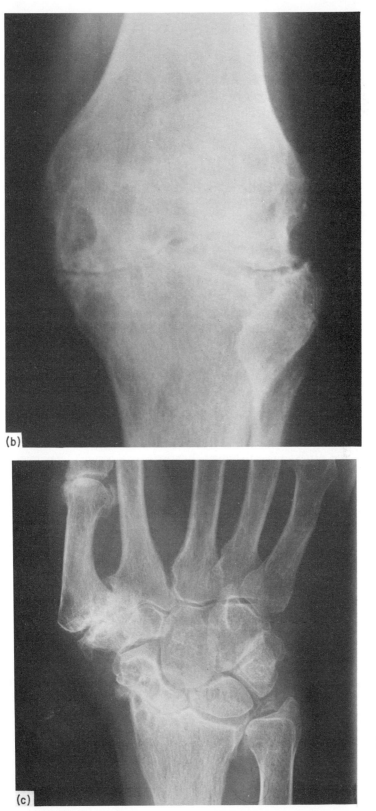

(b)

(c)

183

7.10 Diagnosis

The diagnosis of pyrophosphate deposition may be suspected from clinical findings, radiographs, or synovial fluid findings. It can only be confirmed absolutely if crystallographic analysis of the deposits is carried out.

(a) Clinical diagnosis: any large joint acute arthritis of the elderly may be due to pseudogout; recurrent self-limiting attacks in the elderly are particularly suggestive. Similarly a pattern of severe destructive OA of the knees and wrists in an old person should strongly suggest this disease.

(b) Radiology: shadows of calcific density within joint cartilage raise the possibility of pyrophosphate deposition. The typical sites of involvement are the knee menisci, triangular ligament of the wrist and symphysis pubis. Hyaline cartilage deposits in the absence of fibrocartilage calcification is rare. The typical appearance is of a linear or stippled line of calcification in the mid-zone (Fig. 7.10).

However, there needs to be preservation of joint space and a considerable quantity of crystalline material for these appearances to show up. It is not uncommon for the deposits to be invisible radiographically. Other suspicious features include severe knee disease with extensive patello-femoral involvement and osteophytosis, cystic degenerative changes of the wrist, or severe destructive OA of rapid progression (Fig. 7.11). (The radiological features have been well documented by Resnick, *et al.*, 1977, Martel *et al.*, 1970 and others.)

(c) Synovial fluid: on polarized light microscopy pyrophosphate crystals usually exhibit weak positive birefringence. They are small rectangular crystals that often appear to have a chip out of one corner. They may be few and far between, and difficult to see; a search in fibrin deposits of synovial fluid is worthwhile (see Appendix I).

Further diagnostic help may come from arthroscopy and synovial biopsy. Crystalline deposits may be seen in the synovium or cartilage, visualized under polarized light, or analysed further if they are big enough.

Two difficulties arise in making the diagnosis of pyrophosphate arthropathy:

(1) it is difficult to know when the chondrocalcinosis is a chance finding (some 30% of elderly people have it), and when it has clinical significance;

(2) clinical radiographic and synovial fluid findings can occur separately, or in any combination (Fig. 7.12). Because deposition may be asymptomatic, can be invisible radiographically, and may or may not result in crystals being shed into synovial fluid, any combination of diagnostic clues may be present. One criterion is suspicious, two or more make the diagnosis likely, complete certainty only comes if a deposit can be extracted from the joint and analysed.

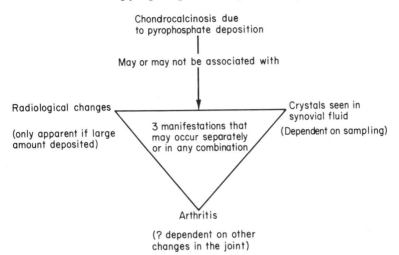

Chondrocalcinosis due
to pyrophosphate deposition

May or may not be associated with

Radiological changes

(only apparent if large
amount deposited)

3 manifestations that
may occur separately
or in any combination

Crystals seen in
synovial fluid
(Dependent on sampling)

Arthritis

(? dependent on other
changes in the joint)

Fig. 7.12   The associations of chondrocalcinosis and their interrelationships.

7.11   Unanswered questions

The questions that remain to be answered are more extensive and fundamental than with gout. Many have been raised in the preceding sections. A few will briefly be re-stated here.

7.11.1   Why fibrocartilage?

Pyrophosphate deposition has a predilection for fibrocartilage; some of the differences between hyaline and fibrocartilage were discussed in Chapter 2, and this selectivity must be a clue to the understanding of the condition.

7.11.2   Why the mid-zone?

Linear deposits in the middle of cartilage are characteristic (Fig. 7.7). This could be due to deposits being favoured at a point midway between the source (chondrocytes?) and sink (synovial fluid?) of pyrophosphate production. Similarly, if calcium diffuses into cartilage from the fluid, it is possible that the critical concentrations of both Ca^{2+} and $P_2O_7^{4-}$ tend to occur midway between the bone and the cartilage surface.

7.11.3   Why do some people get clinical problems?

Crystals can be found in the synovial fluid in the absence of inflammation; similarly, many people with radiological chondrocalcinosis get no cartilage damage. On the other hand, as explained, pseudogout and chronic pyrophosphate arthropathy appear to be distinct clinical entities that only exist in the presence of crystals.

The crystals are therefore a necessary, but not a sufficient, cause for the arthritis. Whether there are differences in the crystals themselves, or what other factors are needed to trigger arthritis, remain unknown.

7.11.4 Pyrophosphate alone?

Many authors have described coexistent pyrophosphate and hydroxyapatite crystals ('mixed crystal deposition'). How often other calcium phosphates occur as well is not clear, but the finding raises several further questions; perhaps hydroxyapatite nucleates the growth of pyrophosphate or vice versa. As will be discussed in the next chapter, even less is known about articular deposition of hydroxyapatite.

7.12 Summary and hypothesis

Crystals of calcium pyrophosphate dihydrate are frequently found in fibrocartilage menisci, and less commonly in hyaline joint cartilage. Increasing age predisposes to this phenomenon, which also occurs in younger people as a result of various metabolic diseases, local abnormalities, or as a familial trait. In some cases no obvious cause is apparent, although a local increase in inorganic pyrophosphate production by chondrocytes probably occurs.

Pyrophosphate deposition may be apparent from radiological examination, or the crystals can be visualized in synovial fluid or biopsy samples. Deposits may be asymptomatic, but can be associated with acute attacks of arthritis, or with chronic destructive joint changes.

The exact relationship between deposits and joint disease is not clear. The hypothesis shown in Fig. 7.9 suggests that a variety of local or general abnormalities can predispose to pyrophosphate formation, including damage produced by joint instability, osteoarthritis and a variety of other arthropathies. Deposition can reduce cartilage resistance to wear, and crystals may be shed into the joint space initiating acute inflammatory episodes. Destructive changes and crystal shedding are most likely if there is some pre-existing joint damage. Thus pyrophosphate deposition can act as an 'amplification loop' in chronic arthritis, but is less likely to cause damage if associated with ageing. Metabolic or familial causes only result in damage when the deposits have been present for many years.

Further reading

DIEPPE, P.A. (1978) New knowledge of chondrocalcinosis. *J. Clin. Path.* **31,** (Suppl 12), 214.

KOHN, N.N., HUGHES, R.E., MCCARTY, D.J. and FAIRES, J.S. (1962) The significance of calcium phosphate crystals in the synovial fluid of arthritic patients. *Ann. Intern. Med.* **56**, 738.

MCCARTY, D.J., KOHN, N.N. and FAIRES, J.S. (1962) The significance of phosphate crystals in the synovial fluid of arthritic patients: The Pseudogout Syndrome. *Ann. Intern. Med.* **56**, 711.

MCCARTY, D.J. (1976) Calcium pyrophosphate dihydrate deposition disease. *Arth. Rheum.* **19**, 275.

MCCARTY, D.J. (1977) Calcium pyrophosphate dihydrate crystal deposition (pseudogout syndrome): Clinical Aspects. *Clin. Rheum. Dis.* **3**, 61.

RUSSELL, R.G.G (1976) Metabolism of inorganic pyrophosphate. *Arth. Rheum.* **19**, 465.

ZITNAN, D. and SITAJ, J. (1976) Natural course of articular chondrocalcinosis. *Arth. Rheum.* **19**, 363.

Text references

ALEXANDER, G.J.M., DIEPPE, P.A., DOHERTY, M. and SCOTT, D.G.I. (1982) Pyrophosphate arthropathy: a study of metabolic associations and laboratory parameters. *Ann. Rheum. Dis.* (in press)

ATKINS, C.J., MCIVOR, J., SMITH, P.M., HAMILTON, E. and WILLIAMS, R. (1970) Chondrocalcinosis and arthropathy: studies in haemachromatosis and in idiopathic chondrocalcinosis. *Quart. J. Med.* **39**, 71.

BENNETT, R.M., MALL, J.C. and MCCARTY, D.J. (1974) Pseudogout in acute neuropathic arthropathy: a clue to pathogenesis, *Ann. Rheum. Dis.* **33**, 563.

BIRD, H.A., TRIBE, C.R. and BACON, P.A. (1978) Joint hypermobility leading to osteo-arthritis and chondrocalcinosis. *Ann. Rheum. Dis.* **37**, 203.

BJELLE, A. (1981) Cartilage matrix and hereditary pyrophosphate arthropathy. *J. Rheum.* **8**, 959.

BOCHER, J., MANKIN, H.J. and BERK, P.N. (1965) Prevalence of calcified meniscal cartilage in elderly persons. *N. Engl. J. Med.* **272**, 1093.

BROWN, W.E. and GREGORY, T.M. (1976) Calcium pyrophosphate crystals chemistry. *Arth. Rheum.* **19**, 446.

BYWATERS, E.G.L (1979) Lesions of bursae tendons and tendon sheaths. *Clin. Rheum. Dis.* **5**, 883.

CABANEL, G., PHELIP, X. and BERTHAN, M.P. (1969) Chondrocalcinose articulaire diffuse et phenomenones rheumatoides. *Communication* No **667/776** Piestany.

DIEPPE, P.A., ALEXANDER, G.J.M., JONES, H.E., DOHERTY, M., SCOTT, D.G.I., MANHIRE, A. and WATT, I. (1982) Pyrophosphate arthropathy: a clinical and radiological study of 105 cases. *Ann. Rheum. Dis.* **41**, 371.

DOHERTY, M. and DIEPPE, P.A. (1981) Acute pseudogout: crystal shedding or acute crystallisation. *Arth. Rheum.* **24**, 954.

ELLMAN, M.H., BROWN, N.L. and PORAT, A.P. (1980) Laboratory investigations in pseudogout patients and controls. *J. Rheum.* **7**, 77.

ELLMAN, M.H. and LEVIN, B. (1975) Chondrocalcinosis in elderly persons. *Arth. Rheum.* **18**, 43.

GATTER, R.A. and MCCARTY, D.J. (1967) Pathological tissue calcifications in man. *Arch. Path.* **84**, 346.

GRAHAM, R., SUTOR, D.J. and MITCHENER, M.B. (1971) Crystal deposition in hyperparathyroidism. *Ann. Rheum. Dis.* **30,** 597.

HEARN, P.R. and RUSSELL, R.G.G. (1980) Formation of calcium pyrophosphate crystals *in vitro. Ann. Rheum. Dis.* **39,** 222.

HOWELL, D.S., MUNIZ, O.E. and MORALES, S. (1981) in *Epidemiology of Osteoarthritis,* (J. G. Peyron, ed), Geigy Publication, Basel.

HOWELL, D.S., MUNIZ, O.E., PITA, J.C. and ENIS, J.E. (1975) Extrusion of pyrophosphate into extracellular media by osteoarthritic cartilage incubates. *J. Clin. Invest.* **56,** 1473.

LAGIER, R. and BAUD, C.A. (1968) *Pathological calcifications of the locomotor system.* In *IV Symposium European des Tissues Calcifies.* (ed. G. Milhaud), Paris, pp. 109–13.

LUST, G., NUKI, G. and SEEGMILLER, J.E. (1976) Inorganic pyrophosphate and proteoglycan metabolism in cultured human articular chondrocytes and fibroblasts. *Arth. Rheum.* **19,** 479.

MARTEL, W., CHAMPION, C.K., THOMPSON, G.R. and CARTER, T.L. (1970) A roentgenologically distinctive arthropathy in some patients with pseudogout syndrome. *Am. J. Roent. & Nuc. Med.* **109,** 587.

MCCARTY, D.J., HOGAN, J.M. and GATTER, R.A. (1966) Studies on pathological calcifications in human cartilage. *J. Bone Joint Surg.* 48A, 309.

MCCARTY, D.J., SILCOX, D.C., COE, F., JACOBELLI, S., REISS, E., GENANT, A. and ELLMAN, M. (1974) Diseases associated with calcium pyrophosphate dihydrate crystal deposition: a controlled study. *Am. J. Med.* **56,** 704.

MCGUIRE, M.K.B., HEARN, P.R. and RUSSELL, R.G.G. (1980) in *Studies in Joint Disease I.,* (A. Maroudas and E. J. Holborow, eds), Pitman Medical, London.

PRITCHARD, M.H. and JESSOP, J.D. (1977) Chondrocalcinosis in primary hyperparathyroidism. *Ann. Rheum. Dis.* **36,** 146.

RESNICK, D., NIWAYAMA, G., GOEGAN, T.G., UTSINGER, P.D., SCHAPIRO, R.F., HASELWOOD, D.H. and WEISNER, B. (1977) Clinical, radiographic and pathological abnormalities in calcium pyrophosphate dihydrate deposition disease. *Radiology* 122, 1.

RICHARDS, A.J. and HAMILTON, E.B.D. (1974) Destructive arthropathy in chondrocalcinosis articularies. *Ann. Rheum. Dis.* **33,** 196.

SEEGMILLER, J.E., FAIRE, G., NETTER, P., GAICHER, A. and LUST, G. (1981) Elevated intracellular inorganic pyrophosphate in fibroblasts and lymphoblasts cultured from patients with dominantly inherited chondrocalcinosis. (Abstract, 15th International Congress of Rheumatology, June 1981.)

STOCKMAN, A., DARLINGTON, L.G. and SCOTT, J.T. (1980) Frequency of chondrocalcinosis and avascular necrosis of the femoral heads in gout: a controlled study. *Ann. Rheum. Dis.* **39,** 7.

WILKINS, W.E., DIEPPE, P.A., MADDISON, P. and EVISON, G. (1982) Osteoarthritis and articular chondrocalcinosis in the elderly. *Ann. Rheum. Dis.* **40,** 516.

WILSON, D.A. and IRVIN, W.S. (1981) Amyloidosis and chondrocalcinosis. *J. Rheum.* **8,** 355.

WOLKE, K. (1935) Uber meniskas-und gelenkknorpelverkalkungen. *Acta Radiol.* **16,** 577.

ZINN, W.M., CURREY, H.L.F. and LAWRENCE, J.S. (1969) The prevalence of chondrocalcinosis. *Communication* No **779,** Piestany.

Chapter 8

DISEASES OF CALCIUM PHOSPHATE DEPOSITION

8.1 Introduction

The system calcium-phosphate–water can precipitate as a number of different minerals depending on the conditions of temperature, pH and concentration. The main mineral constituent of vertebral bone, hydroxy-apatite $(Ca_5(PO_4)_3(OH))$ seems to be the most stable cystalline form under *in vivo* conditions. Properly situated, as a fine parallel dispersion in the tough collagenous matrix of bone, it is the source of the rigidity of our supporting skeleton. Hydroxyapatite packed tightly with a small amount of organic coating on the crystals gives tooth enamel its ceramic hardness whilst the underlying dentine with more organic matrix provides a tough support for the brittle surface. Thus hydroxyapatite crystals must be maintained in the correct proportions, in the right places and with the right morphology if the animal suspended from this skeletal framework is to survive. Clearly this requires that, at least locally, conditions are maintained such that the crystals neither grow nor dissolve. As will be discussed below, this is achieved by maintaining the body supersaturated with respect to hydroxy-apatite but below the level at which spontaneous precipitation occurs.

From Chapter 3 it can be seen that this means that thermodynamics favour precipitation but the system lacks an easy mechanism. Thus the body is vulnerable to unwanted pathological precipitation of hydroxyapatite, ectopic calcification, whose consequences will be serious in proportion to the functional significance of the tissues involved. Joints are very vulnerable to the effects of pathological calcification both because this may limit joint mobility and because the release of fragments of crystals, bone or cartilage, into the joint space as a result of pathological calcification or accident are likely to initiate inflammation.

In addition to hydroxyapatite there is evidence for other calcium phosphate crystals in pathological joints such as brushite (calcium hydrogen

189

phosphate dihydrate, $CaHPO_4 \cdot 2H_2O$) and possibly calcium phosphate $(Ca_3(PO_4)_2)$ or octacalcium phosphate $(Ca_8H_2(PO_4)_6 \cdot 5H_2O)$.

In this chapter we will discuss the conditions controlling calcium phosphate precipitation *in vivo* and then the pathological disturbances of this system.

8.2 Crystallization of hydroxyapatite

Measurements of the solubility and crystallization behaviour of apatites is made very difficult by the small size of the crystals and their tendency to form mixed phases with the other calcium phosphates (Bachra, 1970). A recent measurement of hydroxyapatite solubility gives a value for the solubility product $((Ca^{2+})^5(OH)((PO_4^{3-})^3)$ as 2×10^{-59} at $37°$ C when corrected for ionic strength effects and formation of soluble complexes (McDowell *et al.*, 1977). Taking the normal calcium and phosphate levels in plasma (10 mg/100 ml Ca^{2+} and 3.4 mg/100 ml P) and allowing for the various bound forms and the equilibria between HPO_4^{2-}, $H_2PO_4^-$ and PO_4^{3-} we obtain a concentration product of about 2×10^{-50}. This translates into a supersaturation of roughly ten-fold. The existence of supersaturation of hydroxyapatite in plasma is reasonable as we have seen before that dissolution is generally an easier process than crystallization. Thus in order to maintain a stable skeletal structure it is necessary that the system remain supersaturated but less concentrated than the metastable limit at which precipitation readily occurs. Unlike the other, disease-causing precipitations we have discussed, we also expect some biological mechanism for locally turning the precipitation process on and off, and for causing local dissolution so that bone formation can be controlled. It is worth noting in this context that mammalian blood also has a 410% supersaturation for calcite (calcium carbonate) precipitation but this process is known to be inhibited by phosphate at the concentrations found in the blood. Carbonate is similarly believed to inhibit phosphate precipitation.

In vitro studies of calcium phosphate precipitation show that unseeded hydroxyapatite does not precipitate directly but at concentrations above about 4 mM calcium and phosphate (16 mg/100 ml Ca^{2+}, 12 mg/100 ml P), an amorphous calcium phosphate precipitates. This phase shows no regular crystal structure by X-ray diffraction and has a calcium:phosphorous atomic ratio of around 1.3:1 to 1.4:1. This is between the 1:1 expected for a $CaHPO_4$ compound and the 1.67:1 of hydroxyapatite. Left in aqueous suspension the precipitate then converts, possibly through an intermediate octacalcium phosphate crystalline phase, to hydroxyapatite and the calcium:phosphorus ratio increases but does not reach the ideal hydroxyapatite value (Nancollas and Tomazic, 1974). Excess phosphate is apparently bound to the crystal surfaces (Fig. 8.1). Moreno *et. al.* (1977; 1981) have

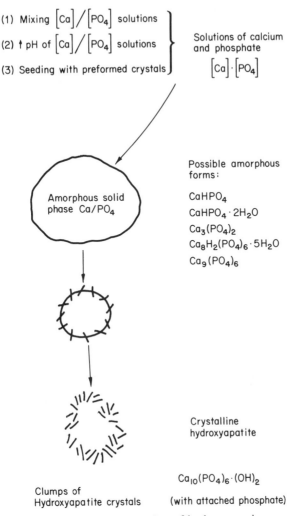

(1) Mixing $[Ca]/[PO_4]$ solutions

(2) ↑ pH of $[Ca]/[PO_4]$ solutions

(3) Seeding with preformed crystals

Solutions of calcium and phosphate

$[Ca]\cdot[PO_4]$

Amorphous solid phase Ca/PO_4

Possible amorphous forms:

$CaHPO_4$

$CaHPO_4\cdot 2H_2O$

$Ca_3(PO_4)_2$

$Ca_8H_2(PO_4)_6\cdot 5H_2O$

$Ca_9(PO_4)_6$

Crystalline hydroxyapatite

Clumps of Hydroxyapatite crystals

$Ca_{10}(PO_4)_6\cdot(OH)_2$

(with attached phosphate)

Fig. 8.1 *In vitro* formation of hydroxyapatite.

recently found that if hydroxyapatite seeds are introduced into a sufficiently dilute solution at physiological pH and temperature, they will grow. The solution must be saturated with respect to hydroxyapatite but not any of the other phases.

Normal *in vivo* bone calcification seems to follow a similar path, thus an area of amorphous 'haze' is seen in the electron microscope ahead of the calcification front. This is believed to be due to precipitated amorphous calcium phosphate which subsequently converts to crystalline hydroxyapatite. However the weight of evidence from X-ray diffraction is now against an initial amorphous precipitate (Glimcher *et al.*, 1981).

The calcification is apparently initiated by matrix vesicles. These minute rounded bodies are formed from small areas of the cell membranes of stem cells which bud off into the extra-cellular area. These consist of a double-layered cell membrane and are rich in calcium, phosphorus and alkaline phosphatase. Minute hydroxyapatite crystals are observed to form on the membrane surface of these vesicles, then form nodules which later coalesce to form the organized sheets of hydroxyapatite which characterize bone.

Electron microscopy shows hydroxyapatite crystals in bone to be plates 5 nm thick by 35 nm wide aligned parallel to the collagen fibres. This relationship with collagen suggests that collagen may have an epitaxial role in hydroxyapatite nucleation and it has been shown that in various *in vivo* and *in vitro* situations collagen fibres can inititate apatite precipitation. However, other collagenous tissues do not normally calcify. Also the initial hydroxyapatite precipitate is random with no orientational relation to the collagen fibres so the role of collagen is probably secondary to the other factors controlling calcification.

The inhibiting effect of carbonate on apatite precipitation was mentioned above. Pyrophosphate is also a very effective inhibitor of mineralization, acting as a crystal growth poison, and may be important in the control of mineralization *in vivo* (Francis, 1969; Russell *et al.*, 1970). Organic phosphonates, stable analogues of pyrophosphate, have a similar effect and are under investigation as drugs for the treatment of pathological mineralization. They can apparently also be used to inhibit bone resorption.

Based on the foregoing, we can develop a picture of normal calcification which may need modification with new research. Body fluids are supersaturated with respect to hydroxyapatite but below the concentration at which amorphous calcium phosphate precipitates; (this, *in vitro*, is a precursor of hydroxyapatite formation). Collagen is capable of promoting hydroxyapatite formation, not necessarily via the amorphous phase, at normal *in vivo* levels of calcium and phosphate, but this effect is opposed by natural levels of pyrophosphate, carbonate and possibly other growth poisons. Calcification is initiated *in vivo* by matrix vesicles which raise local calcium and or phosphate levels to the point where precipitation starts. Possibly pyrophosphate levels are also reduced by alkaline phosphatase so that crystal formation is enhanced.

Bone mineralization is the subject of great research activity at the moment and the reader is advised to refer to books and reviews mentioned at the end of this chapter, and to the journal *Calcified Tissue Research* for more detailed accounts of this fast-moving subject.

8.3   Pathological calcification

From our discussion of hydroxyapatite chemistry it is clear that normal physiological fluids contain calcium and phosphate concentrations that are supersaturated but will not precipitate *in vitro*. Collagen can apparently act as a nucleating agent and will, in isolation, cause precipitation. This tendency is opposed by natural poisons *in vivo*, including pyrophosphate which inhibit the process. Within this framework we can expect calcification in the presence of increased calcium or phosphate levels which might serve to overwhelm the inhibition. Alternatively loss of inhibition, for instance by pyrophosphate decomposition, will also lead to calcification. These effects may arise out of a general metabolic disturbance, out of local changes caused directly by injected substances or by disruption of the mechanism controlling normal, matrix vesicle based, calcification.

As listed in Tables 8.1 and 8.2 a large number of factors are believed to affect the calcification process. Possible activators and inhibitors are listed with the role they are believed to play. However, it must be remembered that an effect seen under relatively simple circumstances *in vitro* may not be reproduced in the presence of many other complex chemical species *in vivo*. Similarly large, non-physiological, changes in concentrations of species can produce an effect where a small change would be ineffective due to the buffering action of the environment.

Table 8.1   Activation of calcification

Possible activators of calcification	Possible role
Matrix vesicles	Source of phosphate, calcium
Collagen	Nucleating agent
Alkaline phosphatase	Destroys pyrophosphate
Matrix components (proteoglycans and others)	Alter calcium concentration
Mitochondria	Concentrate calcium
Fluoride	Stabilize hydroxyapatite
Multivalent metals	Complex pyrophosphate
Electrical cathode	Concentrates calcium? pH changes?

8.3.1   Experimental soft-tissue calcification

Calcification of the soft tissues can be induced in experimental animals by a number of different techniques introduced by Selye and Berczi (1970). The methods fall into two types: (1) those which involve a systemic stimulus with, or without, a further local challenge (calciphylaxis); and (2) those in which soft tissue calcification is induced by a local damage alone (calcergy).

Table 8.2

Inhibitors of calcification	Possible role
Pyrophosphate $(P_2O_7^{4-})$	Growth poison
Diphosphonates $(-P-O-P)$	Growth poison
Magnesium	Binds phosphate
Carbonate	Growth poison
Adenosine triphosphate (ATP)	Source of PP_i
Proteoglycans	Bind calcium
Phospholipids	Bind calcium
Dentine phosphoproteins	Bind phosphate

Calciphylaxis involves sensitization of experimental animals with something that will alter the metabolism of calcium and phosphorus. The most widely used agent has been vitamin D given to rats to raise the serum calcium 24 h prior to local injection of a metallic or other agent; this results in a plaque of calcification appearing at the site of local challenge. A variety of modifications of this model have been described by Selye and others, and it has been shown that in the presence of prior sensitization with vitamin D intravenous metallic compounds will result in spontaneous calcification in the thyroid and elsewhere, and also allow local calcification to be produced by agents such as dextran, which degranulate mast cells, or by other agents which produce local vasodilation. It has been presumed that in a situation where calcium and phosphorus are mobilized in the body fluids, local vasodilation or mast-cell degranulation produces locally higher calcium and/or phosphorus levels in the tissues, thus initiating the process of calcification.

Calcergy is the production of a plaque of calcification locally, without prior sensitization of the experimental animal (Fig. 8.2). A wide variety of different compounds have this action, although the vast majority are salts of divalent or trivalent metals. Potent and commonly used compounds include chromium chloride, lead acetate and potassium permanganate. A local effect on calcium and phosphorus concentrations is thought to be important.

The mechanism of calcification in calciphylaxis and calcergy has been studied by a number of authors. The sequence of pathological events at the local site of damage apparently includes increased cellular activity, with the local production of alkaline phosphatase by fibroblasts, and a decrease in total proteoglycans of the connective-tissue ground substances. This is followed by the appearance of aggregates of amorphous calcium phosphates, and hydroxyapatite-like crystals, often in close association with collagen fibres. Later, more dense coalesced areas of crystalline hydroxyapatite form, and may be surrounded by a macrophage reaction with the production

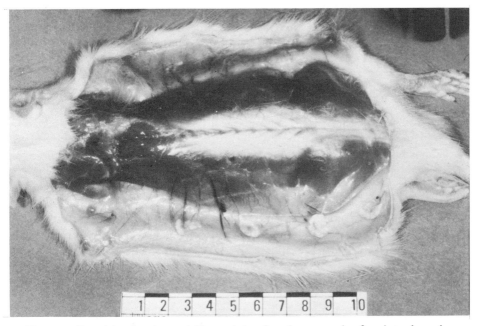

Fig. 8.2 Rat skin showing calcific nodules found one week after intradermal injection of potassium permanganate (calcergy).

of foreign body giant cells in many cases. Intra-cellular calcification is occasionally observed, often in association with mitochondria. In general, cell death seems to follow, rather than precede, the production of the calcification. These experiments, therefore, lend support to the concept that collagen can act as a nucleation site for hydroxyapatite, and that alkaline phophatase has an important role to play in calcification. This may be due to its ability to mobilize phosphate, or to remove the inhibiting effect of pyrophosphate.

The crystals produced by calciphylaxis and calcergy appear to be rod-shaped hydroxyapatite crystals, appearing from an amorphous calcium phosphate solid phase, as observed in the *in vitro* situation. Crystals are rather larger than bone crystals, up to 50 Å wide and 1000 Å long. They coalesce to form plates or nodules of calcified tissue.

Inhibition of these reactions can be achieved in a number of ways which emphasize the interaction of local and general factors in soft-tissue calcification. A low calcium or phosphorus diet will reduce calciphylaxis, and there are strain differences between rats. Paradoxically, the older the rat, the more difficult it is to induce the reaction. A number of agents will induce local inhibition of calciphylaxis and calcergy, including chelating agents, e.g. penicillamine, which would support the importance of the role of free metal in these phenomena. Inhibition can also be induced by diphosphonates and other crystal poisons, and by other large molecules that may affect crystal surfaces (Table 8.2).

8.3.2 Pathological calcification in man

Ectopic calcification occurring in the soft tissues in man is usually divided into metatastic and dystrophic, the former implying a general systemic abnormality predisposing to calcification, and the latter implying a local pathological state alone (Table 8.3). The distinction is strengthened by the fact that the distribution of the calcified tissues tends to be quite different in the two types of condition. However, overlaps do occur. For example, in patients on haemodialysis and with calcific periarthritis the sites of deposition of apatite are often similar to those of patients who appear to have only a local problem in the tendons around the joints. The likely truth is that both local and general factors interact, as exemplified by the models of calciphylaxis and calcergy.

Table 8.3   Ectopic calcification. The formation of relatively insoluble calcium salts in the soft tissues

---

Anatomical descriptions
  (1) Generalized
  (2) Localized
Aetiological descriptions
  (1) Metastatic: secondary to abnormalities in calcium
                  or phosphate metabolism
  (2) Dystrophic: secondary to local tissue damage
  (3) Mixed: due to a combination of (1) and (2)

(a) *Metastic calcification*. A general tendency to soft-tissue calcification occurs in particular in hyperparathyroidism and other pathological conditions leading to hypercalcaemia, and in hypoparathyroidism in which there may be a large increase in levels of serum phosphate. Hypercalcaemia is more commonly the cause than hyperphosphatasia, and inorganic phosphate levels of greater than $8 \, mg/100 \, ml$ are usually necessary before ectopic calcification appears. The sites of involvement in either case are similar, and include the cornea and conjunctivae of the eye, the gastric mucosa, the alveolar septi of the lungs, and blood vessels, in addition to peri-articular soft tissues. The reason for this peculiar distribution is not known, although it has been suggested that these sites tend to have a more alkaline medium, which might predispose to the formation of insoluble calcium phosphates.

Involvement of the eye may be the only visible evidence of ectopic calcification, and slit lamp examination to detect corneal calcification is an important part of the investigation of patients with hypercalcaemia.

Patients on regular haemodialysis become susceptible to generalized soft-tissue calcification, although, as mentioned above, the distribution in this case is often similar to that of people with multiple periarticular calcification, or to that seen in families with heredo-familial joint and arterial calcification (see below). Renal failure and calcification is discussed further in Chapter 10.

(b) *Dystrophic calcification.* Dystrophic calcification is defined as deposition of calcium phosphate mineral in areas of local tissue damage, without obvious systemic cause. It occurs in association with a number of rheumatic diseases, as well as in a wide variety of other pathological conditions. Classification is difficult, although an attempted breakdown of the main types of condition involved is shown in Table 8.4.

Table 8.4   A classification of soft-tissue hydroxyapatite deposition

---

(1) In association with connective tissue diseases
  Scleroderma (calcinosis cutes, calcinosis circumscripta)
  Dermatomyositis (calcinosis universalis)
  Systemic lupus erythematosus
  Relapsing polychondritis
(2) Generalized soft-tissue calcification
  Metastatic: in hypercalcaemia or hyperphosphataemia
  Dystrophic: tumoral calcinosis
       myositis ossificans progressiva
       heredo-familial vascular and articular calcification
(3) Localized soft-tissue calcification
  Asymptomatic calcific periarthritis
  Acute calcific periarthritis
   General variants: Familial
          Polyarticular in haemodialysis
   Local variants: Calcific supraspinatus tendinitis
         Trochanteric syndrome
         'Tennis thumb'
         'Praying stones' (and others)
  Secondary to severe joint damage
        Charcot joints
        Infections
  Secondary to severe muscle damage
        Myositis ossificans
  Neurological diseases
  Diffuse idiopathic skeletal hyperostosis

8.4 Hydroxyapatite deposition and joints

Ectopic calcification can occur in fibrocartilaginous joints, and in or around the synovial joints.

There are two main sites of deposition of apatite involving the synovial joint. They are (1) the *periarticular tendons*, and (2) the *intra-articular cartilage*. Occasionally some systemic predisposition to calcification is present, but in the majority of cases a purely local abnormality appears to be the problem, and calcification often follows other pathological changes. In this sense, articular and periarticular calcification is analogous to the changes that occur following tuberculosis, pancreatitis or other reactions which cause secondary calcification.

8.4.1 The crystals involved

X-ray diffraction and analytical electron microscopy have been carried out on deposits from small numbers of patients. The material obtained is generally amorphous-looking in the polarized light microscope and the individual particles are too small to be clearly identified. Small, rounded globules may be seen in fluid of tissues extracted from inflamed areas, and, as described below, sections of tissue which becomes calcified will show the areas of mineral formation with the standard stains. However, examination of the crystals themselves can only be done using electron microscope or diffraction techniques. The actual crystals and mineral bodies observed in periarticular and articular calcification take on several rather different morphological forms (Fig. 8.3). These include:

(i) small crystals with a calcium–phosphorus ratio consistent with that of hydroxyapatite. These would appear to be rather larger than those found in bone, varying from 0.01 to 1 μm in length (8.3(c));

(ii) small nodules, apparently consisting of clumps of needle-shaped hydroxyapatite crystals. They are observed both in articular cartilage and synovial fluid and are usually in the order of 0.1 to 1 μm diameter (8.3(d)).

(iii) rather larger ovoid smooth-surfaced bodies which can vary in size from 0.1 to 5 μm diameter, which also have the characteristic calcium–phosphorus ratio of hydroxyapatite, but in which definite crystals cannot be seen (8.3(a));

(iv) several authors have also described a number of other amorphous or crystalline looking particles, varying in size and shape, with calcium–phosphorus ratios on electron microscopy which are equivalent of, or near to, that which one would expect to get from that of apatite. Whether these indeed consist predominantly of hydroxyapatite or a similar mineral, or whether they are other forms of amorphous calcium orthophosphate remains unclear.

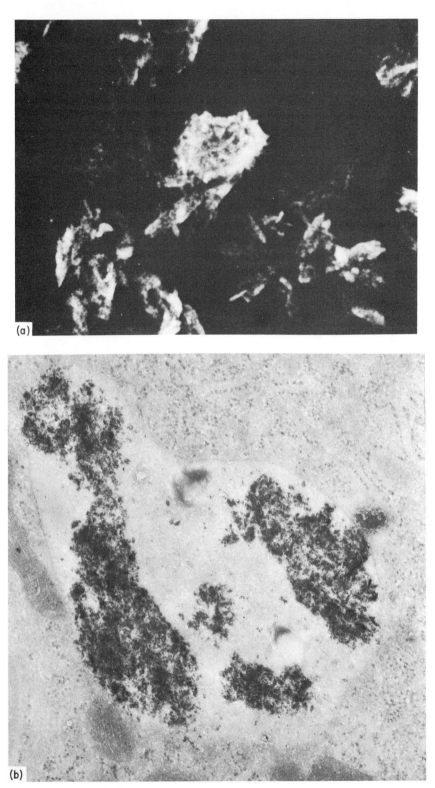

Fig. 8.3 *See overleaf.*

Fig. 8.3 Transmission and scanning electron micrographs of hydroxyapatite crystals in joints. (a) Scanning electron micrograph of nodules of hydroxyapatite crystals found in the synovial fluid in osteoarthritis (× 64 000). (b) Hydroxyapatite crystals in a phagolysosome of a synovial fluid polymorphonuclear leucocyte in acute synovitis (× 96 000). (c) Hydroxyapatite crystals extracted from an area of periarticular calcification (× 120 000). (d) Nodules of hydroxyapatite in articular cartilage (× 25 000).

In the light microscope, the only particles likely to be seen are the larger ovoid bodies (c). These have most frequently been observed in the thick inflammatory fluid aspirated from areas of acute periarthritis. Because of the difficulty in observing and analysing the mineral in synovial or periarticular fluid as hydroxyapatite, few studies have been done. Where diffraction has been carried out, it has invariably shown a pattern characteristic of hydroxyapatite.

8.4.2   The pathology of apatite deposition

As already mentioned, apatite deposition involving the synovial joints occurs primarily in two sites, the tendons and the cartilage. The pathological and ultra-structural findings in each case are quite different.

In *tendons*, it is generally accepted that calcification follows some other pathological reactions. Calcification of tendons is relatively common. Radiographic surveys of the population give figures that vary from 2 to 7% and suggest that the commonest sites are the shoulder (in the rotator cuff), and the hip (around the greater trochanter). The condition appears to be slightly commoner in men than in women, and is most often seen in the sixth decade. The structure of a tendon, and some of the pathological reactions involving calcification that can occur are shown in Fig. 8.4. This shows the various zones of the tendon attachment to muscle and bone. The collagen fibres of the tendon, merge with the cartilaginous and osseous zone at the point of attachment (the enthesis). The blood supply to the tendon comes from the muscle above and from the periostium of bone below, and also from its surrounding membrane. Where a lot of friction is involved, a further membranous structure, the tendon sheath, forms. The osseous zone at the insertion into the bone can extend, forming bony spurs and tendon calcification. Furthermore, in some tendons, small bony areas normally occur (sesamoid bones) and can be thought of as a technique for strengthening the tendon at that point. Finally, and of interest here, small areas of calcification, varying in size and extent, can form in the middle of the tendon later in life. Pathological examination of such tendons appears to show some alteration in matrix staining, and the presence of active cells, which may be increased in number and activity. Deposits have frequently been observed in relatively avascular or traumatized areas, and it is suggested that the change in the vascular supply, or other damage, may cause alteration in cellular function or metaplasia of the cells, resulting in a milieu with less inhibition to calcification than is normally the case. Deposits generally remain contained within the tendon, but may leak out from this site and shower into the surrounding tendon sheath or burase, where an inflammatory reaction may be set up (Fig. 8.4).

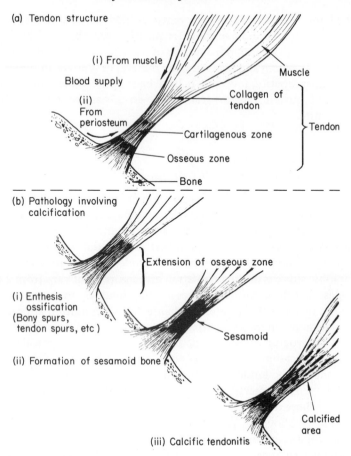

(a) Tendon structure

(i) From muscle

Blood supply

(ii) From periosteum

Muscle

Collagen of tendon

Cartilagenous zone

Osseous zone

Tendon

Bone

(b) Pathology involving calcification

Extension of osseous zone

(i) Enthesis ossification (Bony spurs, tendon spurs, etc)

(ii) Formation of sesamoid bone

Sesamoid

Calcified area

(iii) Calcific tendonitis

In *articular cartilage* a quite different pathological association of calcification has been observed. Where apatite is deposited in hyaline cartilage, it is usually in smaller amounts than in calcified tendons, and in the mid and lower zones of the cartilage, near to the bone–cartilage junction, (Fig. 8.5). Using high-resolution analytical electron microscopy, Ali and colleague (1978, 1981) observed calcifying matrix vesicles in the mid zone of abnormal (osteoarthritic) hyaline articular cartilage. These vesicles are similar to those seen in normal bone calcification and form in the peri-chondrocyte zone (Ali *et al.*, 1970). Crystals of hydroxyapatite first form in or on the membrane of these bodies, and further growth of crystals results in small nodules of dense mineral about 0.1–0.5 μm diameter. Extension of the ossified zone of the cartilage also occurs in osteoarthritis, and small plate-like apatite crystals have been observed on the abnormal surface in this disease, by Ali and Griffiths (1981).

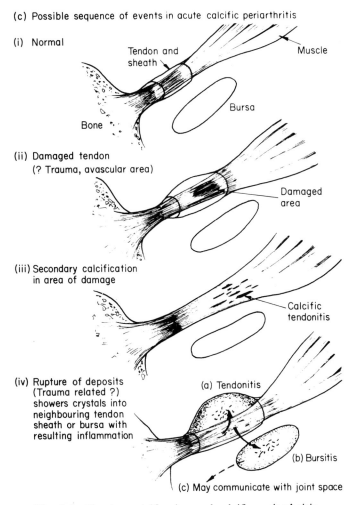

(c) Possible sequence of events in acute calcific periarthritis

(i) Normal
Tendon and sheath
Muscle
Bursa
Bone

(ii) Damaged tendon
(? Trauma, avascular area)
Damaged area

(iii) Secondary calcification in area of damage
Calcific tendonitis

(iv) Rupture of deposits (Trauma related ?) showers crystals into neighbouring tendon sheath or bursa with resulting inflammation
(a) Tendonitis
(b) Bursitis
(c) May communicate with joint space

Fig. 8.4   Tendon calcification and calcific periarthritis.

Synovial calcification can be seen in a variety of disorders, and can apparently result from at least three different mechanisms. These are (1) metaplasia of the cells to form cartilagenous areas which secondarily calcify and sometimes ossify, as in osteochondromatosis; (2) bone fragments taken up by the synovium, having been released from damaged bone or osteophytes; and (3) phagocytosed or acquired microcrystals of apatite coming to the synovium from cartilage via synovial fluid. Synovial calcification has been studied by Doyle (1979) and his colleagues, particularly in the context of osteoarthritis. Mineral deposits are certainly

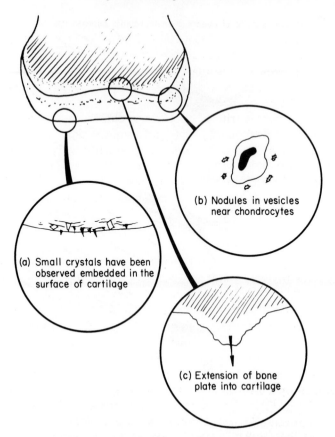

Fig. 8.5 Calicification of articular cartilage. Three different pathological findings involving hydroxyapatite formation in articular cartilage.

frequently seen, but bear no obvious relationship to other pathological reactions in the synovium, including inflammatory changes.

Articular calcification is associated clinically with both acute and chronic joint disease, as are urate and pyrophosphate deposition. Thus hydroxyapatite crystals have been found in the synovial fluid (and synovium) of patients with acute synovitis and chronic osteoarthritis. However, the number of patients studied is small and few controlled series are available.

8.4.3   Acute inflammation associated with articular and periarticular apatite deposition

Although most periarticular calcification probably remains asymptomatic, a well-defined acute inflammatory syndrome can be associated with it; this is acute calcific periarthritis (Canner and Decker, 1964; Pinals and Short, 1966). The clinical associations of intra-articular apatite depositions are much less clear, and although an acute synovitis and osteoarthritis are both possible accompaniments, the lack of large surveys detailing the presence or absence of this mineral in a variety of different conditions make the interpretation of this work difficult.

Acute calcific periarthritis is a distinct clinical syndrome. It most commonly affects the shoulder, but other frequent sites include the toes, fingers, wrists, hips and knees; and as in other acute crystal-induced syndromes multiple sites may be involved, although the majority are monoarticular. The condition is thought to follow the release from a preformed deposit in the tendon, as shown in Fig. 8.4. Support for this hypothesis comes from the frequent observation that the clearly defined area of periarticular calcification seen on the X-ray disappears during the acute attack. The clinical syndrome can be quite dramatic, with a fairly sudden onset of severe pain, stiffness, swelling and redness of the area, often following minor trauma. The very severe pain may rival that of gout, and left untreated will slowly subside over a period of days or even weeks. If the area is aspirated, a small amount of thick white opalescent fluid is extracted from a tendon sheath or bursa. It is rich in polymorphonuclear leucocytes and small rounded globules of mineral may be seen by light microscopy. Powder diffraction studies of spun deposits show the presence of hydroxyapatite. Therapy is similar to that of pseudogout. It may respond to colchicine or anti-inflammatory drugs, and does well with aspiration and local injection of steroids. Surgical excision is occasionally carried out, and said to be a successful way of terminating the acute attack. Hydroxyapatite is known to be capable of causing acute inflammation like pyrophosphate and urate crystals. It is, therefore, thought that acute calcific periarthritis is analogous to acute gout or pseudogout, and a good example of an acute crystal-induced inflammatory disease. It would appear that in some cases, the crystals find their way from associated bursae, and from the peritendon area, into the joint space itself of a nearby synovial joint, resulting in acute synovitis in conjunction with the periarthritis (Fig. 8.4, Table 8.5).

The majority of cases of calcific tendonitis are sporadic and idiopathic (Table 8.6). Occasional associations include generalized disorders predisposing to calcification, such as hyperparathyroidism, vitamin D excess, and patients on haemodialysis. Occasional familial cases, with premature multiple sites of involvement have been described and the condition also appears to occur with an increased frequency in the presence of ideopathic skeletal hyperostosis, although confirmation of this is still needed. The characteristic radiographic appearance is usually the clue to diagnosis, if an acute inflammatory attack occurs secondarily to deposition.

Several groups have reported the presence of nodules or microcrystals of

Table 8.5 Deposition of calcium orthophosphates in periarticular and articular tissue

---

(1) Probable major mineral content:
   Microcrystalline hydroxyapatite
(2) Probable primary sites of deposition:
   (a) Periarticular tendons
   (b) Intra-articular hyaline cartilage
(3) Associated clinical conditions:
   (i) Tendon calcification:
       (asymptomatic in the majority)
       Acute calcific periarthritis
       calcific tendinitis
       calcific bursitis
       (synovitis)
   (ii) Cartilage calcification:
       Acute synovitis
       Osteoarthritis

Table 8.6 Calcific periarthritis

---

(1) Occurs in 2 to 3% normal population
   Commonest in:
       men
       sixth decade
(2) Commonest sites
       shoulder (rotator cuff)
       hip (around greater trochanter)
(3) Usually sporadic and idiopathic
(4) Occasional associations:
       hyperparathyroidism
       vitamin D excess
       haemodyalysis
       familial cases
       idiopathic skeletal hyperostosis

---

hydroxyapatite in the synovial fluid of inflamed joints. Using analytical electron microscopy, Schumacher *et al.* (1979) have detected microcrystals in a few cases of idiopathic acute synovitis, and in patients with osteoarthritis. Using a totally different technique, involving a binding to hydroxyapatite of radio-labelled diphosphonate, Halverson and McCarty (1979) have reported the presence of hydroxyapatite in osteoarthritic fluids, but not in other diseases, and found some correlation between the severity of osteoarthritis and the amount of mineral. In our own studies, a series of patients with rheumatoid arthritis, gout, pyrophosphate arthropathy, osteoarthritis and miscellaneous forms of synovitis have been studied using analytical electron microscopy. As in Schumacher's studies, we found a few cases of ideopatic synovitis in which hydroxyapatite crystals were identified in the absence of any other obvious cause of the condition. The only other type of fluid in which nodules or crystals of hydroxyapatite were found was osteoarthritis, and some mineral was detected in about a third of the cases investigated (Dieppe, 1981).

These crystals and nodules are very similar to those seen in calcifying articular cartilage. It is presumed that they form in the cartilage and are released through clefts into the synovial fluid and may be picked up by the synovium. This escape of crystals into the joint space apparently then triggers an inflammatory response.

Thus hydroxyapatite crystals have been found in cases of acute synovitis, with or without osteoarthritis, in which there is no other apparent cause for the inflammation. In view of the known phlogistic potential of the particles, and by analogy with calcific periarthritis, it is assumed that the crystals cause the synovitis. However, hydroxyapatite synovitis without osteoarthritis or calcific periarthritis is probably a rare cause of arthritis.

8.4.4   Chronic destructive arthritis in association with hydroxyapatite crystals

The provocative finding of these crystals in the synovial fluid in people with osteoarthritis raised the possibility that the chronic arthropathy equivalent to chronic tophaceous gout, or chronic pyrophosphate arthropathy, was osteoarthritis (Huskisson *et al.*, 1979; Dieppe, 1981). This important idea is discussed in the next section.

8.5   Relationship between apatite deposition and osteoarthritis

Crystalline monosodium urate monohydrate, calcium pyrophosphate and hydroxyapatite are each associated with an acute inflammatory disease, and with a chronic destructive form of arthritis. In the case of urate crystals the acute inflammation of the attack of gout is more striking and better known that the chronic destructive joint disease which can occur with continued deposition of the salt. In the case of calcium pyrophosphate dihydrate crystals, acute inflammatory attacks of pseudogout occur in less than half the patients who have arthritis in association with deposits. However, the vast majority have a chronic destructive arthritis similar to osteoarthritis, but differing in its distribution and radiographic features. In the case of hydroxyapatite one acute inflammatory condition is calcific periarthritis, a well established but rare disease. It seems likely that occasional cases of acute synovitis are also related to the presence of this crystal. The finding that hydroxyapatite crystals are often seen in patients with osteoarthritis raises questions as to the role of the crystals in this important disease.

Osteoarthritis is a disease characterized by destruction of articular cartilage, with increased activity and remodelling of the subchondral bone. There is a variable, mild inflammatory synovitis. Early changes in cartilage include an increase in water content and activity of chondrocytes; late changes involve loss of proteoglycans and disruption of the collagen network, and cell death. Although most authors have assumed that cartilage is the primary target of the disease, the site and nature of the initial lesion is unknown, and it is not clear what causes either the inflammation or the symptoms.

Several authors have described the presence of aggregates of hydroxy-apatite crystals in synovial fluid in osteoarthritis, these were present in 9 of 34 consecutive fluids studied with analytical electron microscopy in our series and in a similar proportion of patients in series described by Schumacher *et al.* (1981) and Halverson and McCarty (1979). Ali has described the formation of hydroxyapatite in chondrocyte matrix vesicles in the middle and lower zones of articular cartilage in osteoarthritis (Fig. 8.5). The numbers of patients studied in the cartilage and fluid analyses are small, and there are insufficient controls (normal patients of the same age). However, histological studies of synovium also show hydroxyapatite aggregates, and it does seem likely that osteoarthritis is associated with deposition of these crystals in many cases.

The possible relationships between crystal deposition and cartilage damage have been discussed in the context of pyrophosphate in Chapter 7, and are shown schematically in Fig. 8.6. Some change in the metabolism or structure of the cartilage could lead to secondary crystal deposition; the finding of matrix vesicle mineralization in osteoarthritis suggests that a change in chondrocyte activity to a growth phase might occur.

Even if the crystals are a secondary phenomenon, there is some reason to believe that they may contribute to further damage. The deposits in cartilage are likely to make the tissue more brittle, and surface crystals could cause increased wear. Small apatite crystals embedded in the superficial parts of fibrillated articular cartilage have recently been described, supporting this argument. Hydroxyapatite crystals can also cause a synovitis in experi-mental animals, and could therefore contribute to the inflammation of osteoarthritis. However, we have not been able to find any close association between clinical and thermographic evidence of inflammation and the presence of crystals; and it seems likely that there are many other causes of this aspect of the disease.

Some recent clinical and experimental studies have highlighted a new

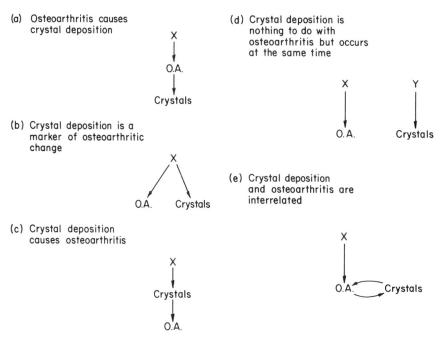

(a) Osteoarthritis causes crystal deposition

(b) Crystal deposition is a marker of osteoarthritic change

(c) Crystal deposition causes osteoarthritis

(d) Crystal deposition is nothing to do with osteoarthritis but occurs at the same time

(e) Crystal deposition and osteoarthritis are interrelated

Fig. 8.6 Possible interrelationships of crystals and osteoarthritis (OA).

approach to the relationship between crystals and chronic destructive joint diseases. Several authors, including ourselves, have noted that hydroxy-apatite crystals are most common in severely damaged joints, without there being much evidence of inflammation. This could imply that deposition is secondary, but McCarty and colleagues (1981) have shown that some of these joints contain high levels of activated collagenase and proteases, and that these destructive enzymes can be released from synoviocytes in culture stimulated by hydroxyapatite crystals. Similarly, Hasselbacher and colleagues find that cultured synovium can engulf the crystals, resulting in release of destructive and bone-resorbing factors including PGE_2. These reactions could proceed in the absence of overt inflammation. A new cycle of destructive changes, involving hydroxyapatite crystals, has therefore been proposed (Fig. 8.7). Many laboratories are now involved in this area of research and rapid progress can be expected. Destructive and erosive forms of arthritis associated with hydroxyapatite crystals have been reported, and may be due to mechanisms of this sort (Dieppe et al., 1978; Schumacher et al., 1981).

Whether there is enough apatite deposition in osteoarthritis to have a significant effect is not clear. The dose as well as the type of crystal is

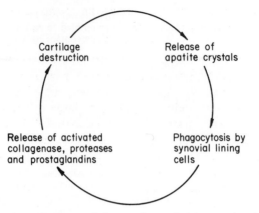

Fig. 8.7 Proposed mechanisms of destructive arthritis associated with hydroxy-apatite crystals (after McCarty *et al.*, 1981).

probably important. However, although they are probably a secondary event in the disease, crystals may accelerate damage in at least some cases – perhaps operating as an 'amplification loop' as suggested for pyrophosphate (Chapter 7).

8.6 Ectopic calcification in other rheumatic diseases

Soft-tissue calcification occurs in many conditions, including several rheumatic diseases (see Table 8.4).

8.6.1 Calcification in the connective tissue diseases

Calcification is sometimes a feature of the connective tissue diseases, in particular scleroderma and dermatomyositis. It is occasionally seen in systemic lupus erythematosus, and cartilage calcification can follow the inflammation of relapsing polychondritis. Previous descriptions of skin calcification included division into calcinosis cutis, and calcinosis universalis. Many cases of the former are now thought to be sclerodermatous, and many of the latter to have been in association with dermatomyositis.

In scleroderma, small deposits of hydroxyapatite are commonly seen in the sub-cutaneous tissues and skin of the hands. These are often asymptomatic, and only apparent radiologically. Sometimes they occur in the pulps of the finger ends, in association with resorption of the tufts of the phalanges, and may ulcerate through to contribute to areas of ischaemic ulceration of the fingers, complicated by the Raynaud's phenomenon (the C.R.S.T. syndrome – calcinosis, Raynaud's phenomenon, sclerodactyly and telangectasia). The severity of these deposits has been said by some authors to parallel the duration of the severity of the skin disease. However, of the most severe cases we have seen many have had minimal other evidence of scleroderma. In addition to the small skin deposits described, these rare patients develop larger nodules, containing large quantities of hydroxyapatite, with a distribution strikingly similar to that of rheumatoid nodules, gouty tophi or xanthelasma. They form nodules, therefore, on the elbows, over the hands, knees and in the Achilles tendons. These nodules look superficially like gouty tophi, and are also similar in that they occasionally become acutely inflamed. They often ulcerate, exuding either masses of hard chalky material, or less commonly, a toothpaste-like substance, particularly if inflammation is present at the time of rupture. They can be extremely painful, and although symptoms wax and wane, the condition tends to be slowly progressive and distressing, and until now, almost untreatable.

In dermatomyositis, calcification is sometimes much more extensive, affecting sub-cutaneous tissues and the connective tissue of muscles. Widespread sheets of calcification may form, particularly in the pelvic and shoulder girdles. Slowly developing areas of large masses of calcification, following remission of childhood dermatomyositis, may develop quite asymptomatically. However, as in the scleroderma deposits, ulceration through the skin can occur.

Similar soft-tissue calcification has also been described in rare cases of systemic lupus erythematosus, and in patients with overlap forms of connective tissue disease. Calcification of the cartilage of the ear has also been described, following episodes of relapsing polychondritis at that site.

The studies of the histological nature of the deposits in these diseases has shown that relatively large hexagonal crystals of hydroxyapatite form, in association with some amorphous deposits of calcium phosphate, as described previously in this chapter. Some authors have suggested that altered collagen and elastin fibres stimulate precipitation at the site of the primary deposit. On the basis of electron microscopy it is also suspected that a secondary chronic granulomatous inflammatory reaction can result from the presence of the crystals. Occasional areas of polymorph infiltration, presumed to be in association with the dispersal of small inflammatory crystals are sometimes observed. The pathology is therefore similar to that described for calcific periarthritis in the previous section; the distribution of these deposits, and their association with connective tissue diseases also strengthens the supposition that ectopic calcification can result from an alteration in connective tissue, leading to loss of the normal inhibition to crystal deposition, or to the appearance of nucleating agents.

8.6.2   Calcification of muscles

Muscle injury may be followed by calcification, and later ossification, of the traumatized area. This is particularly common in the thigh muscles, in association with sporting injury where large haematomas form. The quadriceps apparatus is particularly vulnerable. Within a few days of injury pain and loss of function of the muscle becomes severe. Calcification becomes radiologically apparent within the first few weeks of the injury. Widespread sheets of deposit may then occur at a remarkably rapid rate. The release of large quantities of lipid material into the area has been postulated as being important in these injuries, although it has also been suggested that damage to the periostium, releasing cells with osteogenic potential, is necessary prior to development of this rare, but important, complication of soft-tissue injury.

Muscular calcification also occurs in association with dermatomyositis, and in a rare disorder of young people called myositis ossificans progressiva. This disease is characterized by progressive calcification of the connective tissues in the muscles; it is often associated with congenital skeletal abnormalities, or a family history, suggesting the possibility that inherited defects of connective tissues underlie the cause. Acute pain and inflammation may be seen in the deposits as they first form, although it is not clear whether the crystals are the cause or the result of the inflammatory disease.

8.6.3   Calcification of inter-vertebral discs

Isolated calcification of one or more inter-vertebral discs, particularly in the thoraco-lumbar region, is a not uncommon finding in adults, and has been described in mummies and skeletons from times of antiquity. It is occasionally due to a local pathological abnormality, such as tuberculosis, but more commonly no obvious cause is found. Widespread severe disc calcification raises the possibility of ochronosis, as well as of hypercalcaemia. In a case of ochronosis, the calcification appears to be secondary to the metabolic defect that leads to the build up of abnormal amino acids in the cartilage. This is an interesting and intriguing example of the way in which a single metabolic defect can lead to marked changes in articular cartilage, and to secondary calcification. A specific but rare syndrome involving calcification of the cervical inter-vertebral discs is described in children. It is associated with acute episodes of pain, followed by a disappearance of calcification reminiscent of acute calcific periarthritis. The cause of the deposition in these cases remains unknown.

8.6.4   Costal cartilage calcification

Calcification of the costal cartilages is common in elderly people. Severe calcification in these areas in young people, especially women, is an occasional chance finding. The significance of this is not apparent, and it does not appear to be related to any other ectopic calcium phosphate deposition.

8.6.5   Tumeral calcinosis

This is a rare specific disease, otherwise known as lipocalcinosis granulomatosis, and calcifing collagenolysis. It is characterized by large multi-lobulated calcific masses in the soft tissues, particularly overlying large joints, and perhaps starting in bursae. It is most common at the elbows and hips. It usually occurs in young people, and there is often a familial predisposition. Calcific masses are painless, but recur if removed surgically. Histological examinations show multi-lobulated areas of fluid containing clusters of hydroxyapatite crystals, surrounded by a fibrous capsule and by a mono-nuclear and giant cell reaction. It has been suggested that this condition is associated with metaplasia of connective tissue cells, leading to an attempted formation of ectopic bone tissue. It is associated with a marked increase in local vascularity, which is unusual in these diseases with ectopic calcification, many of which are thought to be associated with ischaemic changes.

8.6.6   Heredo-familial vascular and articular calcification

This is an inherited disorder that has been described by Sharp (1954). In his review of the literature and cases he describes a condition of widespread calcification of the vessels, joints, and periarticular soft tissues. This is another condition whose existence supports the concept that generalized abnormalities of the connective tissue can be responsible for ectopic calcium phosphate deposition.

8.6.7   Periarticular calcification

Calcific periarthritis has already been described. As mentioned, it is now thought that the majority of these cases are due to a local abnormality in tendons or bursae. However, the existence of multiple sites occurring in families argues for a more generalized defect. A large number of regional variants of this phenomenon have been described, many of which have appealing names such as praying stones (calcified masses forming in the infra-patellar or supra-patellar pouches due to frequent kneeling!), tennis thumb, trochanteric syndrome and others. Similar deposits may be seen in patients on long-term haemodialysis, or occasionally in association with a general systemic hypercalcaemia or hyperphosphatasia.

8.6.8   Articular calcification

The articular deposition of hydroxyapatite as a cause of acute synovitis, and in association with osteoarthritis, was described above. It is worth noting here that in addition to these conditions, calcification can occur in and around joints as a secondary phenomenon following severe joint damage from another disease, as in septic joints or in Charcot's joints. This again emphasizes the way in which calcification can be an innocent secondary phenomenon, arising from a wide variety of causes.

8.6.9   Arterial calcification

The formation of crystals of hydroxyapatite is an important feature of arterial disease. To this extent it can be said in a book of this sort that even arterio-sclerosis could, and perhaps should, be viewed as a crystal deposition disease. Subintimal plaques of calcification occur in atheroma, and may be secondary to the appearance of lipid. This is probably a fairly universal phenomenon, and small early areas of change have been described in the arteries from children. A quite different type of hydroxyapatite deposition can also occur in the media of vessels, as in Monkeberg's arteriosclerosis. Deposition can also be a feature of venous disease, and varicose veins. Gravitational ulcers associated with varicose veins may also become calcified.

8.6.10   Traumatic calcification

In addition to myositis ossificans, other haematomas and traumatic areas occasionally become calcified. This is most commonly seen after abdominal injuries such as splenic haematoma, which may turn into a calcified mass. More rarely, traumatic areas in the skin and sub-cutaneous tissue can also become calcified, and calcification at these sites can occasionally be initiated by therapeutic injections.

8.6.11   Calcification following infection

The pathological changes produced by a number of infectious diseases can also lead to calcification; the best known of these is tuberculosis, although the reasons for this being a particular cause of apatite deposition remain unknown, and relatively unexplored. Other infective lesions that can calcify include parasitic cysts, and lungs damaged by chicken pox.

There are therefore a larger number of diverse conditions that can lead to secondary soft-tissue calcification. Small areas of skin taken by punch biopsy and kept in a culture medium will calcify spontaneously if kept for long enough, and experiments in our laboratory suggest that the skin from patients with scleroderma does this much more quickly than that of normal people. This suggests that the tendency to calcify is present in us all, and that the normal inhibitory mechanisms tend to be broken down by diseases affecting connective tissues. Ectopic calcification, as explained in this chapter, can therefore be thought of as a secondary phenomenon due to loss of local inhibitory mechanisms.

8.7   Other calcium phosphates

Much recent attention has been focused on the presence of hydroxyapatite in synovial joints. However, many workers have remarked on the presence of other particles containing calcium and phosphate in joint fluid, and in pathological sections. So far, only brushite has been positively identified.

8.7.1   Brushite (calcium hydrogen phosphate dihydrate $CaHPO_4 \cdot 2H_2O$)

Calcium hydrogen phosphate dihydrate has been identified by X-ray diffraction in studies of pathological human articular and extra-articular calcifications. Gatter and McCarty (1967), using X-ray diffraction, found it was less frequent than pyrophosphate, but more frequently present than hydroxyapatite in meniscal cartilage, and similar findings have been reported by others. However, until recently, cases had not been reported in which brushite crystals were incriminated as a cause of human arthritis.

In the last few years there have been a small number of isolated case reports of a chronic destructive arthritis, punctuated by acute inflammatory episodes, in which X-ray diffraction analysis has suggested that brushite is present either alone or in conjunction with pyrophosphate. It has therefore been suggested that these crystals should be added to the list of causes of chondrocalcinosis and of acute and chronic arthritis. However, problems arise in this identification for at least two reasons. First, unless synovial fluid or other tissues are collected under oil to prevent the escape of CO_2 and alteration in pH, brushite is liable to crystallize out of solution. We have observed a number of samples in which crystals of brushite appeared in synovial fluid left on the bench, where no crystals have been found on immediate examination (Fig. 8.8). Second, brushite has many similarities to pyrophosphate. Thus they both form positively birefringent needle- or monoclinic-shaped crystals which could be confused in the polarized light microscope, and on analytical electron microscopy the calcium–phosphorus ratios are almost identical. Further work, using other techniques such as diffraction, will therefore be required to establish the frequency of the crystal, and great care will have to be taken in avoiding its formation *in vitro*.

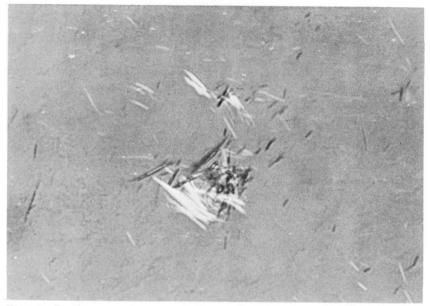

Fig.8.8   Brushite crystals forming in synovial fluid after its removal from the joint; note the star-shaped deposit suggesting *in vitro* rather than *in vivo* formation (× 600).

8.7.2   *Other calcium phosphates*

Use of the analytical electron microscope has revealed a large number of other particles in joint fluid. Many of these contain calcium and phosphorus, and some have a crystalline-looking structure. They may represent intermediary forms in the calcium orthophosphate chain, such as tribasic calcium phosphate or octacalcium phosphate. Alternatively, they may be particles of bony origin, or hydroxyapatite or pyrophosphate covered with some other organic material, altering the physical properties and resulting in inaccurate calcium–phosphate ratio recordings. However, the frequency with which other particles of this sort are seen, both in the electron microscope and light microscope, strongly suggests that a range of other calcium phosphates remain to be identified in joint tissues. In patients with renal disease others have been identified (see Chapter 10).

8.7.3   *Mixed crystal deposition*

Finally, mention should be made of the presence of mixtures of different calcium phosphate minerals occurring in the same joint. We have described a small number of cases in which there is strong evidence for the presence of both newly formed hydroxyapatite and pyrophosphate crystals in the same joint, and similar findings have now been reported by Halverson and McCarty (1979). In our small series, these patients have tended to have the more severe destructive forms of osteoarthritis, although insufficient numbers have been identified to establish whether they represent a definite subset (Dieppe *et al.*, 1979). The finding of mixed deposits of crystals strengthens the likelihood that a range of calcium phosphates can be deposited in articular tissue, and also further illustrates the apparent tendency of the joint to mineralize in the pathological state.

8.8 Summary

The hydroxyapatite deposition which is observed in joints can be compared with a number of forms of calcification including experimental model systems, the passive calcification of necrotic tissue *in vivo* or *in vitro*, and normal bone growth. These can be thought of as falling into two categories, those where there is evidence of matrix vesicle involvement and those where there is none.

In the experimental models and in calcification of necrotic tissue, there is no evidence for matrix vesicle action. The calcification appears to be a consequence of increases in calcium or phosphate levels, locally or generally, or of loss of natural inhibition. Alkaline phosphatase would thus act by breakdown of pyrophosphate, and metal ions by complexation with pyrophosphate. Chelating agents may in turn bind the metal ions or interact directly with the crystal surface. Other crystal poisons as well as pyrophosphate may also be important in these reactions.

In normal bone growth and in pathological cartilage, calcification by matrix vesicles is seen, suggesting that the process is under some higher, cellular and hormonal control. The phospholipid membrane of the vesicles may help nucleate crystals, and the high levels of alkaline phophatase in the vesicle can decrease levels of inhibitory pyrophosphate and raise the local phosphate concentration.

Deposition of hydroxyapatite in or around the joints is common. It is often symptomless, but in common with other crystal deposition arthropathies, may be associated with acute inflammation, or chronic destructive changes. Some of the radiographic changes seen are shown in Fig. 8.9. Hydroxyapatite-associated arthropathies are the newest and least well described of the major crystal-related joint diseases; however, there is considerable current interest in the group, and rapid progress can be expected.

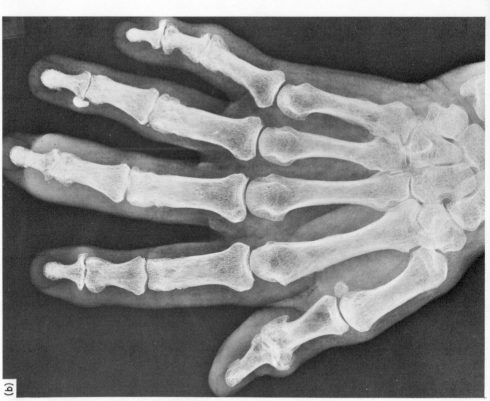

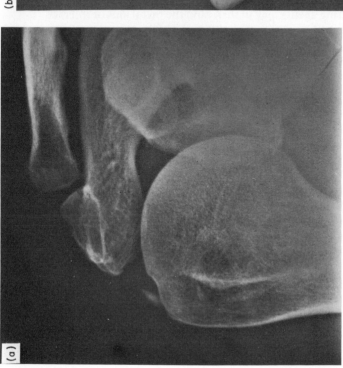

Fig. 8.9 Radiographs of hydroxyapatite deposition in and around the joints. (a) Calcific periarthritis of the shoulder joint. The speck of non-trabeculated calcification between the head of the humerus and acromioclavicular joint is caused by a deposit of hydroxy-apatite in the supraspinatus tendon. (b) Calcific periarthritis in the hand. The deposit of calcific density around the ring finger distal inter-phalangeal joint was asymptomatic. A similar deposit in the adjacent middle finger ruptured, setting off an acute inflammatory response; the soft tissue swelling is shown.

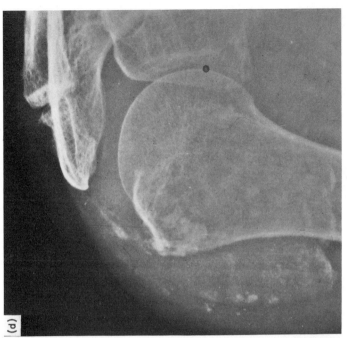

(d)

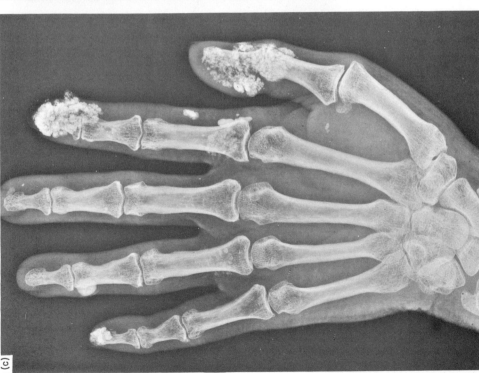

(c)

Fig. 8.9 (c) Calcinosis cutis in scleroderma. The calcific deposits around the fingers, due to hydroxyapatite deposition in the subcutaneous tissue, are often seen in this disease, and are unusually extensive in this case. (d) Synovial calcification of the shoulder joint in a patient with a large effusion rich in hydroxyapatite crystals.

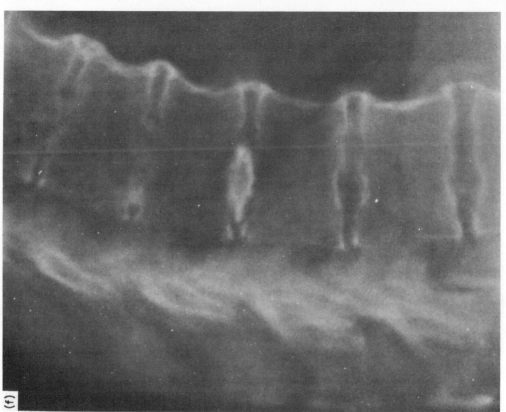

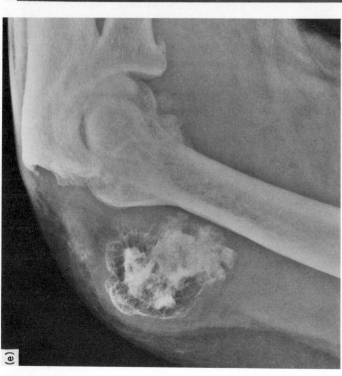

Fig. 8.9 (e) Synovial calcification of the elbow in a case of synovial osteochondromatosis. In this disease the calcification is associated with metaplasia of synovial cells and the formation of ectopic cartilage deposits, some of which calcify and may ossify. (f) Lateral tomograph of the thoracic spine showing bridging osteophytes anteriorly, and calcification of an intervertebral disc.

Further reading

BLUMENTHAL, N.C. and POSNER, A.S. (1973) Hydroxyapatite: mechanism of formation and properties. *Calc. Tiss. Res.* **13**, 253.

DIEPPE, P.A. (1979) Crystal deposition disease and the soft tissues. *Clin. Rheum. Dis.* **5**, 807.

NUKI, G. (ed) (1981) *The aetiopathogenesis of osteoarthritis*, Pitman Medical, London.

ROBERTSON, W.G. (1973) Factors affecting the precipitation of calcium phosphate *in vitro*. *Calc. Tiss. Res.* **11**, 311.

SELYE, H. (1962) *Calciphylaxis*, University Chicago Press, Chicago, Illinois.

Text references

ALI, S.Y. (1978) New knowledge of osteoarthritis. *J. Clin. Path. (suppl)* **12**, 191.

ALI, S.Y. and GRIFFITHS, S. (1981) New types of calcium phosphate crystals in arthritic cartilage. *Sem. Arth. Rheum.* **11**, (Suppl 1), 124.

ALI, S.Y., SAJDERA, S.W. and ANDERSON, H.C. (1970) Isolation and characterisation of calcifying matrix vesicles from epiphyseal cartilage. *Proc. Nat. Acad. Sci.* **67**, 1513.

BACHRA, B.N. (1970) Calcification of connective tissue. *Conn. Tiss. Res.* **5**, 165.

CANNER, J.E.Z. and DECKER, J.L. (1964) Recurrent acute arthritis in chronic renal failure. *Am. J. Med.* **36**, 571.

DIEPPE, P.A. (1981) Inflammation in osteoarthritis and the role of micro-crystals. *Sem. Arth. Rheum.* **11** (Suppl 1), 121.

DIEPPE, P.A., DOYLE, D.V., HUSKISSON, E.C., WILLOUGHBY, D.A. and CROCKER, P.R. (1978) Mixed crystal deposition disease and osteoarthritis. *Br. Med. J.* **1**, 150.

DOYLE, D.V., HUSKISSON, E.C. and WILLOUGHBY, D.A. (1979) A histological study of inflammation in osteoarthritis: the role of calcium phosphate crystals deposition. *Ann. Rheum. Dis.* **38**, 192.

FRANCIS, M.D. (1969) The inhibition of calcium hydroxyapatite crystal growth by polyphosphonates and polyphosphates. *Calc. Tiss. Res.* **3**, 151.

GATTER, R.A. and McCARTY, D.J. (1967) Pathological tissue calcifications in man. *Arch. Path.* **84**, 346.

GLIMCHER, M.J., BONAR, L.C., GRYNPAS, M.D. *et al.* (1981) Recent studies of bone mineral: Is the amorphous calcium phosphate theory valid? *J. Crystal Growth* **53**, 100.

HALVERSON, P.B. and McCARTY, D.J. (1979) Identification of hydroxyapatite crystals in synovial fluid. *Arth. Rheum.* **22**, 389.

HASSELBACHER, P. (1981) Microcrystalline material stimulates secretion of collagenase and PGE_2 by synovial fibroblasts. *J. Rheum.* **8**, 1013.

HUSKISSON, E.C., DIEPPE, P.A., TUCKER, A. and CANNELL, L. (1979) Another look at osteoarthritis. *Ann. Rheum. Dis.* **38**, 423.

McCARTY, D.J., HALVERSON, P.B., CARRERA, G.F., BREWER, B.J. and KOZIN, F. (1981) Milwaukee shoulder—association of microspheroids containing hydroxyapatite crystals, active collagenase, and neutral protease with rotator cuff defects. *Arth. Rheum.* **24,** 464.

McDOWELL, H., GREGORY, T.M. and BROWNE, W.E. (1977) Solubility of $Ca_5(PO_4)_3OH$ in the system $Ca(OH)_2–H_3PO_4–H_2O$ at 5, 15, 25 and 37° C. *J. Res. Nat. Bur. Stand.* **81A,** 273.

MORENO, E.C. and VARUGHESE, K. (1981) Crystal growth of calcium apatites from dilute solutions. *J. Crystal Growth* **53,** 20.

MORENO, E.C., ZAHRADNIK, R.T., GLAZMAN, A. and HWU, R. (1977) Precipitation of hydroxyapatite from dilute solution upon seeding. *Calc. Tiss. Res.* **24,** 47.

NANCOLLAS, G.H. and TOMAZIC, B. (1974) Growth of calcium phosphate on hydroxyapatite crystals: effect of supersaturation and ionic medium. *J. Physical Chem.* **78,** 2218.

PINALS, R.S. and SHORT, C.L. (1966) Calcific periarthritis involving multiple sites. *Arth. Rheum.* **9,** 566.

RUSSELL, R.G.G., MUHLBAUER, R.C., BISAZ, S., WILLIAMS, D.A. and FLEISCH, H. (1970) The influence of pyrophosphate, condensed phosphate, phosphonates and other phosphate compounds on the dissolution of hydroxy-apatite *in vitro* and in bone resorption induced by parathyroid hormone in tissue culture and in thyroparathyroidectomised rats. *Calc. Tiss. Res.* **6,** 183.

SHARP, J. (1954) Heredo-familial vascular and articular calcification. *Ann. Rheum. Dis.* **13,** 15.

SCHUMACHER, H.R., MILLER, J.L., LUDVICO, C. and JESSAR, R.A. (1981) Erosive arthritis associated with apatite crystal deposition. *Arth. Rheum.* **24,** 31.

SCHUMACHER, H.R., SOMLYO, A.P. and TSE, R.L. (1979) Arthritis associated with apatite crystals. *Ann. Intern. Med.* **87,** 411.

SELYE, H. and BERCZI, I. (1970) The present status of calciphylaxis and calcergy. *Clin. Orthop. Rel. Res.* **69,** 28.

Chapter 9

MISCELLANEOUS CRYSTALS AND PARTICLES

Urate, pyrophosphate and hydroxyapatite are the main crystal species found in joints. However, other crystals and particles are occasionally identified in articular tissue. They are of three types: extrinsic particles introduced from outside; intrinsic crystals deposited *in vivo*; and bone and cartilage fragments from the joint itself. Many of these particles are capable of damaging the joints, and are therefore of clinical relevance. The reaction of a joint to the introduction of a foreign body is an interesting model which also throws further light on the pathogenesis of the crystal-induced ar-thropathies. These miscellaneous crystals and other particles are the subject of this chapter.

9.1 Crystals deposited in synovial joints

Examination of synovial fluid in a polarized light microscope will occasionally reveal the presence of small birefringent particles, which are soluble in lipid or other organic solvents. This suggests that organic matter other than that described in the last three chapters, can be deposited in joint tissue, and crystalline forms of both lipids and proteins can appear from time to time. Because of their relative ease of identification, cholesterol crystals stand out in this context, and have been described by a number of authors.

9.1.1 *Cholesterol crystals in synovial fluid*

In 1964, Zuckner and his colleagues described a number of cases of rheumatoid arthritis in whose synovial fluid cholesterol crystals were seen and positively identified by infra-red spectrophotometry. Following that, a number of other authors have reported that occasional effusions in rheumatoid arthritis are milky in appearance, due to the presence of large numbers of cholesterol crystals; more recently, they have also been described in other conditions, including osteoarthritis.

Cholesterol crystals are characteristic, and easy to identify in the polarized light microscope. They form flat, pleomorphic plates, usually with a crosswire surface marking, and often with a chip out of one corner (Fig. 9.1). When present (which is rare) their numbers are often large, and some may be seen inside the leucocytes. They can be further identified either by infra-red spectrophotometry of a synovial fluid pellet, or less directly, by applying a cholesterol solvent such as acetone to the slide containing the synovial fluid.

Cholesterol crystals injected into the synovial joints of animals, or into the rat foot pad or pleural space, cause a small inflammatory reaction prior to their removal; it has therefore been suggested that they may occasionally exacerbate an inflammatory reaction in human joint diseases. It has also been suggested that they may be responsible for some of the joint manifestations of patients with hyperlipidaemias. Patients with both type 2 and type 4 hyperlipidaemia sometimes get transient joint effusions. However, examination of the synovial fluid in these cases has not revealed

Fig. 9.1　Cholesterol crystals in synovial fluid (× 600).

any evidence of crystal formation, and another explanation of the synovitis seems more likely.

The cause of deposition of cholesterol crystals is unclear. The cholesterol may be derived from cell membranes in the effusions, as well as from blood. Crystallisation is most often found in chronic cysts and other long-standing effusions. Presumably enzymatic splitting of cholesterol from cholesteryl esters such as oleate prediposes to deposition. The crystals may be an incidental finding in joint disease, of little clinical significance. An alternative viewpoint has recently been put forward by Fam *et al.* (1981) and Pritzker *et al.*, (1981) who suggest that cholesterol crystals are potent causes of fibrosis, capable of making an important contribution to synovial pathology. Further experimental and chemical studies would seem to be worthwhile.

9.1.2 Cysteine

A case of cysteinosis has been described in which there were joint symptoms and synovitis. Cysteine crystals were found elsewhere, and it was suggested that they might be causing the joint disease. However, cysteine crystals were not, and never have been, found in joints. Thus cysteine is of interest in being one type of crystal deposited in man which does *not* prefer joint tissue to other sites.

9.1.3 Others

In Chapter 6 the possibility was discussed that a variety of purine derivatives other than urates could be deposited. Similarly Chapter 8 mentions the case for a wide variety of other calcium phosphates crystallizing *in vivo*. Other lipids, proteins and salts may occasionally form in joint tissue. A wide range of particles, including calcium phosphates and calcium oxalate have been found in patients with renal failure (Chapter 10).

9.2 Extrinsic crystals and particles found in synovial joints

So far in this chapter, we have considered crystalline particles that form *in vivo* from metabolic products. As outlined in Chapter 2, crystals and other hard particles may also be introduced into the body from outside, and then induce a secondary pathological reaction. In this section, a brief description of diseases induced by these extrinsic particles is given. These particles can be sub-classified into two sorts as shown in Table 9.1. The first sort of particle is introduced by doctors during procedures aiding the diagnosis and treatment of joint disease. The second type are particles entering the synovial joint through the skin during a traumatic episode. As shown in the table, a variety of particles may enter in one of these two ways. The ones which assume the most importance are considered in more detail below, and

Table 9.1 Non-crystalline particles that can be found in joints

(1) 'Iatrogenic'	Fragments of surgical implants
	Bone cement
	Talc, suture material, or other surgical material
	Corticosteroids
	Radioactive colloids
	Radio-opaque dyes
(2) Derived from the joint	Bone fragments
	Cartilage fragments
	Fibrin
	'Rice bodies'
(3) Foreign bodies	Plant thorns and other sharp, penetrating objects
	Grit, dirt and other traumatically acquired foreign bodies

include steroid crystals, fragments derived from surgical implants, and plant thorns.

9.2.1   Iatrogenic disease

During the investigation and treatment of joint disease, a number of procedures may be used involving the injection or surgical entry of foreign materials to the synovial cavity. It is perhaps not surprising that foreign particles introduced in this way can be a cause of joint disease. Investigative procedures include injection of radio-opaque dyes to produce arthrograms, although the material used sometimes produces a transient synovitis. Injections of radioactive colloids are used to produce radiation synovectomy, and they, too, can produce a transient increase in inflammation. Corticosteroid preparations are commonly used to induce a temporary anti-inflammatory effect; in order to prolong the duration of action, crystalline forms are used as these stay within the joint cavity for a longer period. A transient inflammatory reaction maximal for 24–48 h after injection, occurs in a few cases. Surgery can result in talc and other foreign material entering the cavity and inducing secondary damage. Similarly surgical implants and artificial joints may result in metal and plastic particles and fragments of the cement being released into the cavity. Of these various materials, the two best investigated are corticosteroid crystals and the particles coming from surgical implants.

9.2.2   Corticosteroid crystals

The majority of long-acting steroid preparations used for injection into the joint cavity to produce an anti-inflammatory effect are crystalline. These are visible under polarized light microscopy, and may cause diagnostic confusion with other crystals. They may also be responsible for the transient synovitis sometimes seen following these injections.

The appearance of corticosteroid crystals in synovial fluid has been described by Kahn and his colleagues (1970). They found that methyl prednisolone acetate and triamcinalone acetonide produce strongly birefringent crystal fragments that often clump together, whereas triamcinalone hexacetonide and betamethasone acetate produce negatively birefringent rods of varying sizes that could be confused with urate crystals (Fig. 9.2). Incubation with synovial fluid did not alter the crystal appearance, and clinical studies suggest that the crystal may be visible for as long as three months after an injection. Enquiry about previous steroid injections should always be made when birefringent material is observed in a diagnostic tap.

Corticosteroid ester crystals are phagocytosed by cells lining the synovium or within the synovial fluid. Injections into the rat pleural space can produce a transient increase in fluid and cell content prior to the onset of the anti-inflammatory effect of the steroid. Therefore the post-injection flare, beginning several hours after an injection and subsiding after 24–72 h, may be a crystal-induced synovitis. The incidence of post-injection synovitis is

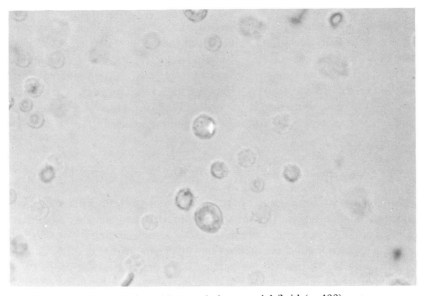

Fig. 9.2   Steroid crystals in synovial fluid (× 400).

variable, and there are very few controlled or comparative studies. Clinical experience suggests that the response may depend on the batch of a preparation used. This could be related to minor alterations in the size and nature of the crystals in the suspension. Clinically and pathologically the effect is of no real significance, but patients should always be warned of the possibility of a flare after an injection.

9.2.3. Fragments from surgical implants

The soft tissues and synovial fluid of joints with a surgical implant have been found to contain many fragments of plastic or metal derived from the material of the implant itself. These have been studied in relation to wear of the implant surface, but are not very important. These particles could set up a low-grade inflammatory reaction, contributing to the late loosening or painful reaction to implants. Areas of granulomatous inflammation around particles derived from metal implants are analogous to other foreign body granulomatous reactions. Immunological responses to the materials are relatively uncommon, but may sometimes contribute to loosening of the prosthesis.

Fragments of polyethylene and polymethylmethacrylate can be identified by polarized light microscopy, and metal particles by analytical electron microscopy. As with other particles, they may become engulfed by the synovium or synovial fluid phagocytic cells. They are sometimes surrounded by chronic inflammatory reactions, which may extend into the area of bone–prosthesis attachment. These granulomas are common in loose prostheses, and have been described in association with skin-test evidence of metal (e.g. cobalt) sensitivity. Fluid from these loose, particle-filled joints may be a murky brown colour, and light microscopy shows a large amount of cell debris, and thousands of particles. Low-grade chronic infection may also be present. The relationship between loosening, infection and particles is probably complex. There is probably a vicious cycle developing, analogous to that of other particle-associated diseases.

9.3 Traumatically acquired particles in synovial joints

Trauma may result in extrinsic particles of a variety of types entering a joint cavity through the skin. Obviously, this is usually associated with major trauma, and a large amount of damage to the surrounding soft tissues and bones, so that any subsequent reaction is overshadowed by other damage. Occasionally, small penetrating particles can reach the joint cavity in the absence of much other damage, and by far the most interesting examples occur with long thin plant thorns, such as the blackthorn.

9.3.1   Plant thorn synovitis

Several cases have been described in which plant thorns have entered synovial joints resulting in a synovitis, with inflammation occurring around the plant thorn; the condition is cured by the removal of the thorn (Sugerman *et al.*, 1977). Although this is not a crystal-induced disease, it is of interest, as it gives some insight into the way that particles can induce an inflammatory reaction, and is at present being investigated in our laboratory and elsewhere.

The commonest plant thorn to produce this reaction in England is the blackthorn (*Prunus spinus*). This is a perennial shrub, which is found widely in hedges and thickets all over the British Isles, but particularly in the West Midlands. It has large numbers of narrow sharp thorns, several inches long, which can easily penetrate the skin and break off leaving a small fragment in the sub-cutaneous tissue, or, on occasions, in a synovial cavity. In America, the date palm, sectriel palm, *Yucca alifolia* (Spanish Bayonet), and rose thorns have all been described as causing inflammatory synovitis. Palm fronds apparently fall to the ground and produce sharp tips after drying in the sun. If one kneels on the ground, or falls over, the brittle tip of the palm can then penetrate through the skin into the synovial joint, as can the blackthorn in England. Because of the seasonal nature of production of brittle palm fronds, the disease in California apparently clusters in the late winter and spring. Strangely, most other plants which also have long spikey thorns, have not been implicated in this way.

Clinically, the condition is characterized by an acute synovitis which has a latency of from one to several days after the injury. This is followed by a relatively asymptomatic period, and then the development of a chronic relapsing arthritis, which may occur long after the injury has been forgotten. Pathological examination shows evidence of a patchy granulomatous reaction of the synovium, and the plant thorns can be identified at the centre of the lesion. Polarized light microscopy will identify the particles, and PAS staining can be used to pick out the vegetable material on the surface of the thorn (Fig. 9.3). The presence of multinucleate giant cells may cause confusion with other granulomous diseases, including sarcoidosis and tuberculosis. Patients tend to present late, only seeking advice after the joint adjacent to an area of minor trauma has gone on being a problem for several weeks. The thorn aspect of the injury may be long forgotten, and any evidence of skin abrasion healed, so that the condition can be difficult to diagnose. Careful examination of spun deposits from synovial fluids and biopsy material usually reveals the presence of thorns. Treatment has generally been surgical removal of the vegetable material from the joint. This leads to complete resolution of the synovitis.

We have been carrying out experimental work on the inflammatory

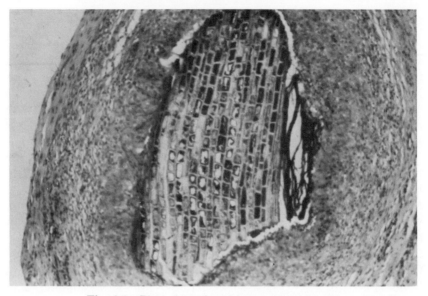

Fig. 9.3 Plant thorn found in synovium (× 40).

reaction to thorns injected into the rat foot pad or intradermally in human forearm skin. In general, thorns have a negative surface charge, similar to that of inflammatory crystals, and the more phlogistic tend to be the more charged. They can also split complement and affect cell membranes, with consequent release of intracellular enzymes. Thorns which are not phagocytosed can still produce a brisk inflammatory reaction, indicating that phagocytosis is not necessarily a feature of particle-induced inflammatory reactions. Why the initial clinical response should subside, to be followed by an intermittent fluctuating synovitis, defies explanation. However, the inflammatory reaction to the particles can vary considerably, and this may depend on the size and surface of the thorn, and on what proteins become attached to its highly charged surface. Further experiments on the thorns may give clues as to the mechanism of crystal-induced synovitis.

9.3.2 Foreign body reactions in joints

A variety of other foreign bodies can occasionally enter joints and cause chronic low-grade inflammatory responses. This is illustrated by a somewhat unusual case seen recently.

The patient was a 22 year-old woman, who was involved in a road traffic accident, during which she suffered minor abrasions to the left knee from hitting the dashboard of her Mini car. She was taken to the local Casualty

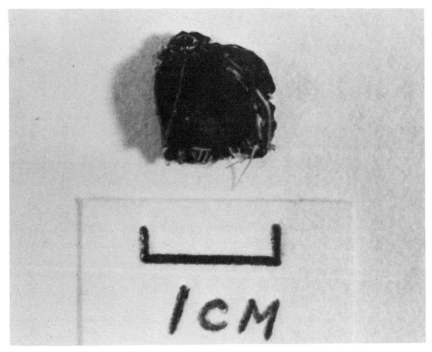

Fig. 9.4 Foreign body found in a joint cavity (see text).

Department, where the small abrasions were cleaned and sutured, and she was discharged. X-rays of the knee showed no bony damage. The pain slowly subsided, and the patient was then well for a short interval, and the injury was forgotten. A few weeks later, however, she started to experience intermittent pain and swelling of the knee, and occasional locking of the knee joint. She repeatedly sought medical advice, but on examination, the knee appeared perfectly normal, and X-rays and haematological investigations were normal. Over the subsequent few months the patient experienced increasing pain, although clinical examination was repeatedly normal, apart from the occasional mild effusion. Advice was sought from a number of different medical practitioners, and the patient was thought to be hysterical; it was suggested that the pain was of psychotic origin. However, in view of her persistent complaints about the joint, and the history of previous trauma, a diagnostic arthrotomy was finally carried out. At operation, a small black foreign body was found in the joint, and a low-grade synovial reaction, but no other abnormality. The foreign body (Fig. 9.4), was examined extensively, and initially defied analysis. However, the use of the infra-red spectrophotometry finally identified the material as Bakelite, derived from the case of the car radio. This is, as far as we know, the first

Fig. 9.5   Scanning electron micrograph of a bone fragment found in synovial fluid (× 1500).

reported case of 'Radio synovitis'. Following arthrotomy and removal of the foreign body, the patient has been perfectly well.

This anecdote illustrates the way in which any foreign body can produce severe pain and a low-grade synovitis, as well as mechanical problems in a synovial joint. It is of interest that, unlike the thorns or crystals, the degree of overt inflammatory reaction caused by this particle was very mild, and a lot of the pain was probably due to the mechanical problems induced by a large foreign body in a moving hinge joint.

9.4   Particles derived from the joint itself

Examination of synovial fluids by light microscopy or ferrography may reveal the presence of numerous fragments of cartilage and bone. These are most frequent in osteoarthritis and destructive joint diseases, and obviously result from the damage occurring at the joint surfaces. However, these particles may also be pathogenic: they may be engulfed by the synovium and initiate low grade inflammation or the release of proteolytic enzymes, and could also act as 'abrasive particles', accelerating surface wear.

In some inflammatory joint diseases, such as rheumatoid arthritis, particles of fibrin, fibronectin, or other insoluble organic matter may accumulate. A possible pathological role of these bodies is also under investigation.

Joint disease thus results in the production of particles derived from the joint itself. These particles may further insult the integrity of the joint. This is yet another example of a vicious cycle, accelerating joint damage, in which particles play an important role (Fig. 9.5).

9.5 Summary

It is clear that a wide variety of crystalline and other particles can produce inflammation and damage in synovial joints. Investigation of the common factors between crystals and the other particles, such as plant thorns, which induce synovitis is underway. It appears that they all have negatively charged surfaces capable of activating proteins such as complement, and interacting with cell membranes. In future we should perhaps think of particle-induced joint disease rather than crystal-related arthropathies.

The joints themselves shed particles into the synovial space. Bone and cartilage fragments are often seen in synovial fluid, especially in osteoarthritis, or in aggressive destructive conditions such as septic arthritis. These particles may induce secondary synovial reactions, as well as having wear properties that may damage the surface of hyaline cartilage. Thus intrinsic joint particles may help accelerate and perpetuate the arthritic process.

Joint particles sometimes initiate arthritis; in other cases they may act as an 'amplification loop', as suggested in the case of pyrophosphate crystals (Chapter 7). Investigation of the particulate content of synovial fluid and cartilage surfaces, by ferrography and other techniques, may advance the understanding of joint diseases considerably.

Further reading

DIEPPE, P.A., CROCKER, P.R., CORKE, C.F., DOYLE, C.V., HUSKISSION, E.C. and WILLOUGHBY, D.A. (1979) Synovial fluid crystals. *Quart. J. Med.* **48,** 533.

EVANS, C.H., MEARS, D.C. and McKNIGHT, J.L. (1981) A preliminary ferrographic survey of the wear particles in human synovial fluid. *Arth. Rheum.* **24,** 912.

GRAHAM, J. and GOLDMAN, J.A. (1978) Fat droplets and synovial fluid leucocytes in traumatic arthritis. *Arth. Rheum.* **21,** 76.

KAHN, C.B., HOLLANDER, J.L., and SCHUMACHER, H.R. (1970) Corticosteroid crystals in synovial fluid. *J. Am. Med. Assoc.* **211,** 807.

Leader Article (1980) Can metal sensitivity loosen joint replacements *Lancet* **ii,** 1284.

Text references

FAM, A.G., PRITZKER, K.P.H., CHENG, P.-T. and LITTLE, A.H. (1981) Cholesterol crystals in osteoarthritic joint effusions, *J. Rheumatol.* **8,** 273.

PRITZKER, K.P.H., FAM, A.G., OMAR, S.A. and GERTZTEIN, S.D. (1981) Experimental cholesterol crystal arthropathy. *J. Rheumatol.* **8,** 281.

SUGERMAN, M., STOBIE, D.G., QUISMORIE, F.P., TERRY, R. and HANSON, V. (1977) Plant thorn synovitis. *Arth. Rheum.* **20,** 1125.

ZUCKNER, J., UDDEN, J., GANTNER, G.E. and DORNER, R.W. (1964) Cholesterol crystals in synovial fluid. *Ann. Intern. Med.* **60,** 436.

Chapter 10

RELATED DEPOSITION DISEASES OF THE KIDNEY

10.1 Introduction

The musculoskeletal system is not the only one to be affected by crystal deposition diseases. There are two principal types of disorder in other systems: (1) ectopic calcification in damaged or dead tissue. This nearly always involves hydroxyapatite, and was discussed in Chapter 8; and (2) stone formation in the ducts of excretory organs such as the kidneys, liver and salivary glands.

The kidney is affected by both these processes; a diseased kidney may calcify (nephrocalcinosis), and renal stones are common. Furthermore, gout and kidney disease are interrelated in several other ways. Patients with long-standing renal failure are particularly prone to crystal deposition, and a wide range of different salts have now been identified in this group. In this chapter parenchymal and calculous diseases of the kidney are considered in the context of crystal deposition diseases of the joints, particularly gout.

10.2 Parenchymal kidney disease

The deposition of hydroxyapatite crystals in the parenchyma of the kidney can occur in hypercalcaemia, or secondary to a wide variety of renal disorders. Therefore nephrocalcinosis can be an example of dystrophic or metastatic calcification (Chapter 8). Urate crystals can also form in the kidney, and the remainder of this section will concentrate on the relationship between gout and the kidney.

10.2.1 Gout and the kidney

Some aspects of the interactions of these two diseases are shown in Table 10.1. If the glomerular filtration rate falls below 20 ml min^{-1}, or urinary flow below 1 ml min^{-1}, serum uric acid levels may rise, predisposing to

234

Table 10.1 Gout and the kidney – interrelationships

(1) Renal failure can cause gout (Glomerular filtration rate <20 ml min^{-1}).

(2) Gout is associated with vascular disease, including nephrosclerosis.

(3) Gouty individuals have abnormal renal glomerular function when compared to normal controls.

(4) Patients with gout are often 'undersecretors' of uric acid, and tend to excrete an acid urine due to a renal tubular abnormality.

(5) A few families with gout and severe premature renal failure have been described.

(6) Between 10 and 20% of gout patients suffer from uric acid renal calculi, which can damage the kidney.

(7) Hyperuricaemia is associated with an increased risk of calcium salt urolithiasis.

gout. Gout secondary to severe renal disease is uncommon, and takes many years to develop. Tophi are not seen, but the disease is otherwise indistinguishable from other types of gout in 'undersecretors'. Gout is associated with hypertension, and an increased risk of cardiovascular disease, so that renal vascular changes are common in gouty patients. Renal glomerular function has been compared in subjects with gout and in age–sex matched controls, and the gout group found to have reduced function that cannot be accounted for fully by hypertension or ageing, suggesting a more specific renal lesion. Tubular function is also abnormal in gout, many patients are undersecretors of uric acid, and tend to excrete an acid urine, with a relatively fixed pH due to a defect in ammonia handling by the kidney. Finally, a few families have been described who have gout and hyperuricaemia in association with premature, severe renal disease, in whom a special mechanism of kidney damage is postulated.

The pathology of the kidney in gout has been studied. Three types of parenchymal change have been observed. Vascular nephrosclerosis is common, and a patchy infiltrate of inflammatory cells with a picture indistinguishable from chronic pyelonephritis can occur. Deposits of monosodium urate monohydrate crystals may be seen, and in some cases an inflammatory reaction surrounds them. Presumably the mechanisms behind the deposition of the crystals, and the resulting damage are similar to those in gout of the joints (see Chapter 6).

Familial gout with renal failure has been described in a few young women. The pathology of the renal lesion was identical to that seen in the typical middle-aged or elderly male with hypertension. This suggests that the lesions described above may be related directly to gout.

Acute uric acid nephropathy is a condition that can occur during the treatment of myeloproliferative diseases, when serum uric acid levels may rise to very high levels. An acid urine, or dehydration increase the risk.

Urinary output falls, and serum creatinine levels rise; there may be a sludge of uric acid crystals in the urine, although the condition is not always easy to diagnose. The tubules become blocked with crystals, and acute tubular necrosis can occur. Treatment should include allopurinol to reduce uric acid production, and maintenance of a high volume of an alkaline urine.

10.3   Urolithiasis

The renal tract is a collecting system for excreted material. The body is therefore deliberately raising the concentration of many solutes in the urine, increasing the risk of crystal formation. Variations in flow, and tube blockage can cause local alterations in pH or concentration locally, and predispose to infection, adding further to the chance of crystalline material forming. It is therefore no surprise that urolithiasis is one of the oldest, and commonest, forms of crystal deposition disease (see Chapter 1).

The types of urinary calculi that grow are shown in Table 10.2. Calcium oxalate with or without calcium phosphates are the most frequent minerals

Table 10.2   Analysis of the composition of renal calculi

Calcium oxalate	39.4%
Calcium oxalate and calcium phosphates	20.2%
Calcium phosphates	13.2%
Magnesium ammonium phosphate	15.4%
Uric acid	8%
Cystine	2.8%
Others	1%

After Westbury (1974).

found. The stones also contain a protein matrix, accounting for about 3% of their weight. Formation depends on the saturation of solute, but is also influenced by urinary inhibitors. As in the cartilage, the body seems to have an inhibitory mechanism to prevent crystal deposition in an area of high risk. The nature and importance of urinary inhibitors is not resolved. Citrate and pyrophosphate are present in urine and act as inhibitors of phosphate precipitation. Calcium oxalate is affected by a high molecular weight factor which is believed to be a specific protein. There is some evidence that recurrent stone formers are deficient in these factors but the case is not proven.

10.3.1   Uric acid stones

Man and the dalmatian coach hound are the only two species to suffer from uric acid urolithiasis. Man excretes about ten times as much uric acid as most mammals, and also tends to form an acid urine, predisposing to crystal formation. The pK_a of uric acid is 5.75, so that the solubility changes dramatically with pH (Fig. 3.11). A relatively acid urine easily becomes superstaturated for uric acid, and it is crystalline uric acid, and not the sodium salt, that is deposited.

Excess metabolic acidity may be absorbed by manufacture of ammonia, or cause reduced urine pH. Many gouty people have an acid urine due to a deficiency of tubular cell formation of ammonia, increasing the risk of stones. Uric acid stones can also form in normouricaemic individuals if they tend to form acid urine, or become dehydrated. These calculi are therefore usually due to hyperuricaemia, hyperuricosuria, dehydration, or an acid urine (Table 10.3). A special cause is ileostomy, which results in a very acid urine, and a high incidence of uric acid stone formation.

Table 10.3 Causes of uric acid calculi

Hyperuricaemia	Acid urine
Hyperuricosuria	(Idiopathic)
Dehydration	

Different surveys give different figures (varying from 5–25%) for the number of people with gout who also get uric acid calculi. The stones are radiotranslucent, relatively small, and often easily passed per urethra. However, they can cause obstruction, and are another possible cause of renal failure in gout (Table 10.1).

10.3.2 Hyperuricosuric calcium urolithiasis

Gout is also associated with an increased incidence of calcium oxalate urinary stones.

Any cause of hyperuricosuria can result in calcium-containing calculi. In this case the urinary pH is usually above 5.5, unlike the acid urine that favours uric acid crystallization.

The cause of the stones is not clear, but two hypotheses have been put forward. Both propose that crystals of sodium urate form in the relatively alkaline urine; one case proposes that these crystals act as epitaxial nucleation sites for calcium oxalate, the other hypothesis suggests that they absorb and inactivate the natural urinary inhibitors of stone formation.

10.4 Crystal deposition in haemodialysis patients

Patients on haemodialysis often develop periarticular calcific deposits, as well as ectopic calcification at other sites. They are also prone to 'gout-like' attacks of arthritis in peripheral joints and to acute calcific periarthritis (Irby *et al.*, 1975). Three main species of crystal have been identified, calcium phosphates, calcium carbonate and more recently calcium oxalate (Irby *et al.*, 1975; Hoffman *et al.*, 1982). Although hydroxyapatite is probably the main constituent of most deposits, carbonated apatite, octacalcium phosphate, calcium orthophosphate and calcium carbonate have all been identified in periarticular tissues. Calcium oxalate has been identified in the synovial fluid of patients with haemodialysis who developed episodes of acute arthritis, and it has been suggested that they may cause the synovitis (Hoffman *et al.*, 1982). This is the only report of intra-articular oxalate deposition to date.

Crystal deposition in this group is largely due to their abnormal metabolic statue; high levels and a $Ca \times P$ product of 70 $(mM)^2$ predisposes to calcium phosphate crystallization, and blood oxalate levels rise as the serum creatinine rises. It is interesting to note that in the face of the systemic metabolic disturbance, it is still the joints and periarticular tissues which are particularly prone to deposition, thus emphasizing the importance of local tissue factors. The sites of deposits are similar to those of idiopathic periarticular and articular hydroxyapatite deposition (Chapter 8).

10.5 Summary

Pathological calcification can occur in any organ of the body, including the heart and the brain. Calculi can occur in all excretory ducts. In nearly all cases the predominant mineral is hydroxyapatite, although gall stones contain cholesterol and bile pigment, urates and oxalate occur in the kidney, and both brushite and urate crystals have been found in salivary gland calculi. Crystal deposition is also a prominent feature of atheroma.

These conditions are beyond the scope of this book, but it is clear that a similar approach, exploring the conditions that lead to solute saturation, crystal growth and tissue damage, can be useful to help understand diseases in many parts of the body.

Further reading

GIBSON, T., SIMMONDS, H.A. and POTTER, C. (1979) Sequential studies of renal function in gout. *Europ. J. Rheum. Inflam.* **3**, 79.

GRAHAME, R. and SCOTT, J.T. (1970) Clinical survey of 354 patients with gout. *Ann. Rheum. Dis.* **29**, 461.

GUTMAN, A.B. and YU, T.F. (1957) Renal function in gout. *Am. J. Med.* **23**, 600.

HOFFMAN, G.S., SCHUMACHER, H.R., PAUL, H. *et al.* (1982) Calcium oxalate microcrystalline-associated arthritis in end-stage renal disease. *Ann. Intern. Med.* **97**, 36.

IRBY, R., EDWARDS, W.M. and GATTER, R.A. (1975) Articular complications of homotransplantation and chronic haemodialysis. *J. Rheum.* **2,** 91.

MALEK, R.S. and BOYCE, W.H. (1977) Observations on the ultrastructure and genesis of urinary calculi. *J. Urol.* **117,** 336.

PERRY, M.C., HOAGLAND, H.C. and WAGNER, R.D. (1976) Uric acid nephropathy. *J. Am. Med. Assoc.* **236,** 961.

SIMMONDS, H.A. (1978) Crystal-induced nephropathy: A Current View. *Eur. J. Rheum. Inflam.* **1,** 86.

SIMMONDS, H.A., WARREN, D.J., CAMERON, J.S., POTTER, C.F. and FAIRBROTHER, D.A. (1980) Familial gout and renal failure in young women. *Clin. Nephrol.* **14,** 176.

YU, T.F. and GUTMAN, A.B. (1967) Uric acid nephrolithiais in gout. *Ann. Intern. Med.* **67,** 1133.

WESTBURY, E.J. (1974) *Br. J. Urol.* **46,** 215.

Chapter 11

TREATMENT

11.1 Introduction

The main aims of treatment in any rheumatic disease include:

(1) relief of symptoms;
(2) maintenance of maximum functional ability;
(3) prevention of further joint damage by the disease process.

A variety of methods are available to attain these goals. They include physical therapy, occupational therapy, education and psychological help, as well as drug treatment and surgery. In this chapter the theoretical and practical aspects of treating acute inflammation and chronic joint damage related to the presence of joint particles will be discussed. The emphasis is on present and future chemotherapeutic avenues.

The drug treatment of joint diseases can be further subdivided into (1) symptomatic therapy, and (2) disease-modifying drugs. Symptomatic drugs include analgesics, which reduce pain perception; anti-inflammatory drugs, which will reduce stiffness and swelling of joints as well as pain; and anti-depressant or other psychotropic agents. None of these drugs have been shown to have any beneficial long-term effect in rheumatic diseases. The other group of agents are those which do influence the long-term outcome. They include gold injections or D-penicillamine, which may induce a remission of rheumatoid arthritis, and allopurinol or uricosuric drugs, which may favourably influence the outcome of gout.

The two chief clinical manifestations of the crystal deposition diseases are acute, self-limiting inflammatory attacks, and chronic destructive changes in the joints. Symptomatic therapy is appropriate for the treatment of the acute episodes, and will be discussed later in this chapter. At the present time, gout is the only crystal-related disease which can be favourably modified by long-term chemotherapy. However, the pathways leading to crystal-induced damage, which have been discussed in the first part of this book can potentially be altered by drugs, and in the first part of the chapter theoretical means of influencing these diseases will be discussed; this should

240

provide a platform for the understanding of any future advances in the therapy of these conditions.

11.2 Chemotherapy of crystal deposition diseases

The main steps involved in the development of a crystal deposition disease are outlined in Fig. 11.1. They include (1) the accumulation of high concentrations of solute at the site of deposition, (2) the nucleation of crystals and subsequent crystal growth, forming (3) a deposit within the tissue. Crystals may be shed from this deposit, helping to initiate the cascade of events that leads to acute inflammation, and the presence of crystal deposits may also lead to chronic destructive changes within the joint. As the figure shows, each pathway in this sequence of events can be modified by chemical agents.

11.2.1 Reducing the concentration of solute

The origin of joint crystals was discussed in Chapter 3. A sufficient concentration of solute is essential before crystal nucleation and growth can occur. One of the most obvious ways of preventing these diseases is to reduce the solute concentration. This can be achieved in a number of ways; the

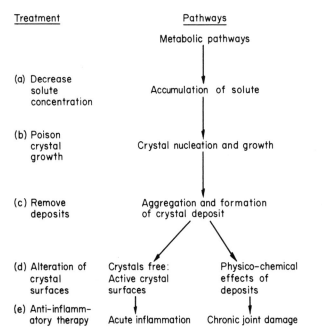

Fig. 11.1   Pathways of crystal deposition diseases and possible sites of therapy.

metabolic pathway producing the solute could be inhibited, or the solute excretion increased; alternatively, the effective concentration could be reduced by introduction of another molecule which will bind to the solute.

The metabolism of uric acid was discussed in Chapter 6, and is particularly amenable to this therapeutic approach. The concentration of serum urate can be reduced by dietary methods, by increasing its excretion through the kidney, or by interfering with production by inhibiting xanthine oxidase (Fig. 11.2). Equivalent therapy is not yet available to treat hydroxyapatite or calcium pyrophosphate dihydrate crystal deposition discussed in Chapters 7 and 8. The concentration of ionized calcium in the body cannot be changed without dire consequences to cell function. Phosphate concentrations are less critical, but are not easy to manipulate without producing problems with bone formation. As yet, the metabolic pathways leading to the accumulation of high concentrations of pyro-phosphate in joint tissues are not fully understood; however, if an enzyme defect such as an excess activity of $5'$ nucleotidase is confirmed (see Chapter 7), it may be possible to find an enzyme inhibitor that will reduce this local excess production and so prevent the formation of pyrophosphate crystals. This is obviously a potential area of exploration in the future, and may produce the 'allopurinol of pseudogout'. Another approach being explored in the case of pyrophosphate deposition is to decrease the effective concentration of $P_2O_7^{4-}$ by increasing the concentration of another ion such

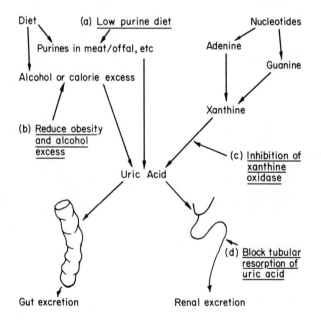

Fig. 11.2   Uric acid metabolism and approaches to achieving hypouricaemia.

as magnesium. Magnesium readily binds to pyrophosphate, producing a highly soluble complex, thus reducing the likelihood of formation of the relatively insoluble calcium pyrophosphate dihydrate.

As explained in Chapter 3, not all joint particles crystallize from solutes *in vivo*. Some (the extrinsic crystals) enter the joint from outside. The equivalent treatment of these cases is easier to understand. One should simply avoid situations which may allow the particles to enter! In practice, this may not be as easy as it sounds. One cannot advise the keen gardener to avoid his rose bushes; nor will the makers of 'sloe gin' be happy to hear that they should not approach the blackthorn for fear of acquiring one of its barbs. In contrast, it should be possible to prevent doctors from putting crystalline steroids into joints and inducing inflammatory flares; hopefully future long-acting steroid therapy will be available in a physical form which is less likely to produce crystal-related synovitis.

11.2.2   Crystal nucleation and growth

Crystal nucleation and growth can be inhibited by crystal poisons: agents that attach to the growth site of a crystal and prevent the accumulation of further molecules (see Chapter 3). Crystal poisons can act at extremely low concentrations to inhibit nucleation of crystals and so increase the supersaturation required to form deposits. When this effect is present, with only a few parts per million of the poison, it is known as a 'threshold effect'. Other crystal poisons will only act effectively at higher concentrations, if their adsorption to the crystal surface is weak. Some apparent cases of crystal poisoning are actually cases of reduction of solute concentration, it is not always easy to distinguish. Poisons may result in the formation of crystals with altered morphology, and/or that contain the entrapped crystal poison.

Naturally occurring crystal poisons within connective tissue and urine have been mentioned, and may include the proteoglycans of articular cartilage, pyrophosphate, and a specific protein factor in urine. Crystal poisons are already being explored for their potential in the treatment of biliary and renal stone formation, and in extra-articular calcification. They have not been widely explored as a treatment of crystal-induced joint diseases, although this is an obvious avenue for future research.

The best known group of crystal poisons are the diphosphonates, whose general structure is shown in Fig. 11.3. These compounds act as crystal poisons to calcium phosphates such as hydroxyapatite. They have been widely used in the treatment of Paget's disease, and trials have also been carried out to assess their effect in a number of extra-articular calcific conditions. Results have generally been disappointing, although if given in

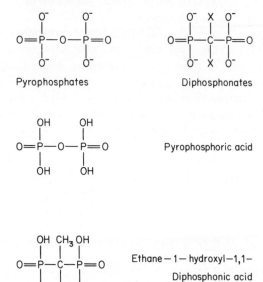

General formulae of pyrophosphates and diphosphonates

Pyrophosphates

Diphosphonates

Pyrophosphoric acid

Ethane − 1 − hydroxyl −1,1−

Diphosphonic acid
(EHDP)

Fig. 11.3 The diphosphonates.

sufficient quantity, prior to the formation of pathological calcification, they can be effective. An example is their use as a prophylactic to reduce the risk of para-articular calcification following hip surgery in patients with ankylosing spondylitis. As might be expected, use of high doses for long periods of time will produce osteopenia due to the inhibitory effect on bone mineralization. These compounds also have a number of actions on cell metabolism, particularly in connective tissue, and their beneficial results in Paget's disease may be due as much to their cellular effects as to their being crystal poisons; they may have modifying effects in other arthritic diseases for the same reason. More is therefore likely to be heard of the diphosphonates in rheumatic diseases, although their actions are only partly those of a crystal poison.

Preliminary experiments are under way in our laboratories and elsewhere with a number of other crystal poisons. Heparin, which is a polymer similar in structure to connective tissue proteoglycan, has been shown to inhibit the formation of urate crystals *in vitro*. A number of other polymers are also being investigated, and some have both threshold effects and retarding effects on crystal growth. An example of the effect of one of these agents is shown in Fig. 11.4. Crystals grown without the threshold agent show typical morphology, while those grown in its presence have an altered

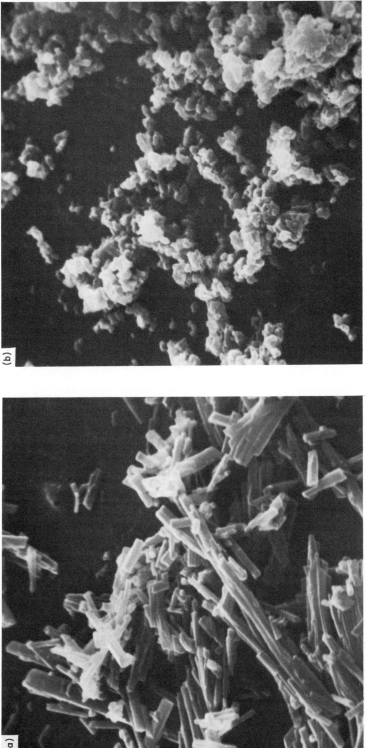

Fig. 11.4 Morphology of monosodium urate crystals grown *in vitro*. Normal crystals (a) are needle shaped and have monoclinic form; those grown in the presence of a poison (b) are rounded, 'amorphous' looking particles (× 900).

shape as well as taking much longer to grow. These agents have also been used in the *in vivo* model of skin calcification of the rat, discussed in Chapter 8. Calcergy can be inhibited by diphosphonates as well as by some of the polymer agents mentioned. If relatively non-toxic agents of this sort can be found, a new avenue for the treatment of crystal-induced joint disease may be opened up.

An alternative approach is to raise the concentration of natural inhibitors. The potential role of pyrophosphate ions in the inhibition of hydroxyapatite formation is mentioned in Chapter 8. It may therefore be useful to raise, rather than lower, the concentration of pyrophosphate in circumstances where hydroxyapatite deposition is favoured. If further understanding of the natural inhibitors in connective tissue is achieved, it may also be possible to enhance their effects, and thus improve the body's natural defence mechanism rather than resorting to the use of other more artifical techniques to prevent the nucleation and growth of crystals. It seems likely that this area will be a fruitful one for future research.

11.2.3 *Removing the crystal deposit*

Once a crystal deposit has formed within the tissue it may still be possible to treat the disease by removing it. If there is a natural turnover of the crystals in the tissue, then this could be achieved by raising the solubility or depressing the dissolved concentration as discussed in Section 11.2.1. Alternatively, physical disruption or removal of the deposit may be possible.

Gouty tophi are sometimes removed surgically, a procedure that needs to be combined with an agent to reduce the likelihood of further crystal deposition. The enzyme uricase has also been used in the treatment of chronic tophaceous gout, in an attempt to solubilize preformed deposits.

Treatment of this sort is again difficult in the case of hydroxyapatite or calcium pyrophosphate deposition. Agents that reduce solute concentrations are not yet available, so that surgical removal of the deposit is likely to be of only temporary benefit. However, pyrophosphate deposition is often confined to knee menisci, and whether removal of the whole fibro-cartilage pad would be useful remains to be seen. If it could be shown that chondrocalcinosis of the knee meniscus is associated with a considerably higher risk than that of removing the joint cartilage, operative removal would be indicated, but this seems very unlikely.

Another approach used in attempts to remove deposits of calcium pyrophosphate dihydrate has been to increase their solubility through the use of magnesium. EDTA and magnesium washouts of the joint produce attacks of pseudogout, perhaps due to crystal shedding associated with solubilization. If this is the case, a combination of further washouts of this sort, plus an anti-inflammatory treatment to cover the acute attack, might result in the removal of the deposit. We have recently attempted joint lavage with high concentrations of magnesium in a small number of cases, with some apparent radiological improvement and without inducing pseudogout.

Physical disruption of deposits could also be attempted. Renal stones have been treated in this way; sonication being used to disrupt the stone, allowing it to be passed through the ureter. However, it seems that such an approach is unlikely to be helpful in joint disease, where disruption of a deposit may well lead to crystal shedding and inflammation, and may also expose an increased surface area for the formation of new crystals, and thus accelerate rather than retard the progress of the disease. It indeed seems likely that the mechanical disruption of crystal deposits within joints is one of the main factors contributing to the susceptibility of this organ to the formation of large numbers of crystals.

11.2.4   Altering crystal surfaces

In Chapter 5, the mechanisms of crystal-induced damage were discussed. The reasons for believing that the charged surface groups of a crystal were important in initiating inflammatory responses were outlined. If this is the case, another potential therapeutic avenue is the neutralization of active surfaces. It seems likely that *in vivo* proteins, such as immunoglobulin, become attached to any free surface, and may be important in initiating inflammation. If a high concentration of an agent which would preferentially bind to crystal surfaces was available locally when the crystals were shed into the joint space, this inflammatory sequence might be prevented. This approach might also be a useful one in research on the mechanisms of crystal-induced inflammation.

11.2.5   Anti-inflammatory drug therapy

Some of the most widely used agents in the treatment of the crystal-induced diseases are the anti-inflammatory drugs. Three types of agent are available. First, colchicine, which is particularly effective in the inhibition of gout, but sometimes helps pseudogout and calcific periarthritis; secondly the non-steroidal anti-inflammatory drugs, which are widely used in most rheumatic diseases; and thirdly steroids, which can be given either systemically or intra-articularly to produce a powerful anti-inflammatory effect. The mode of action and use of these drugs is discussed further in the sections on specific clinical problems. The mode of action is not clearly understood, although it appears to be at a cellular level, inhibiting either the movement of cells into the inflamed tissue, or the release of mediators from phagocytic cells, thus inhibiting the cycle of events that leads to the acute inflammatory response.

11.2.6   Reducing chronic joint damage

At the present time, so little is understood about the mechanisms of chronic joint destruction in relation to the presence of crystals, that very few ways in which the process can be prevented are obvious. However, as further understanding of joint function, joint integrity, and their alteration by crystal deposits is accumulated, advances in this area may also develop.

11.3   The treatment of specific diseases

The remainder of this chapter deals briefly with practical treatment of some of the specific conditions mentioned in earlier parts of the book. Acute inflammation due to joint particles is generally treated with anti-inflammatory drugs, and by removal of the inflammatory fluid containing the particles. The acute conditions are naturally self-limiting, and treatment does not present a significant problem. Treatment of calcific deposits, and of chronic destructive arthropathies, is much more difficult and unsatisfactory. Relief of symptoms and attempts to maintain function are often all that can be offered.

11.3.1   Acute gout

Gout is an extremely painful condition, but a very satisfactory one to treat. The attacks are naturally self-limiting, and the patient is always grateful!

Prevention is better than cure, and approaches to the long-term management of gout are presented below. If, however, the acute attack has already begun, treatment will be demanded. If the patient is already on long-term chemotherapy this should not be altered. Hypouricaemic drugs should not be started during an acute attack, as they may prolong it or make it worse. Aspirin should also be avoided. The joint should be rested, and an anti-inflammatory drug started as soon as possible (Table 11.1). Colchicine given in an initial loading dose of 1 mg followed by 0.5 mg orally every two hours is an effective treatment, although it often produces diarrhoea (thus making the patient run before he can walk!). Colchicine is sometimes recommended as a therapeutic test for gout, but this is unsatisfactory, as pseudogout and calcific periarthritis sometimes respond to the drug as well. Intravenous colchicine is recommended in some centres, and can produce a rapid resolution of gouty inflammation. An alternative anti-inflammatory drug is indomethacin, given in a 'decrescendo' regime, with a loading dose of 75–100 mg, followed by 50 mg four-hourly until the attack begins to

Table 11.1   Treatment of gout

---

(1) Acute attacks
    (i) Avoid aspirin, allopurinol and pure uricosuric drugs.
    (ii) Rest the joint.
    (iii) Non-steroidal anti-inflammatory drugs in initial high dose, tailing off quickly, e.g. indomethacin 100 mg loading dose, then 50 mg q.d.s. decreasing as pain subsides *or* azapropazone 2400 mg first day, then 1200 mg day^{-1}.
OR   (iv) Colchicine 1.0 mg loading dose followed by 0.5 mg two-hourly until side effects develop or attacks subsides.
    In resistant cases intra-articular steroids or systemic ACTH.

(2) Chronic gout
    (i) Remove any obvious cause, drugs, obesity etc.
    (ii) Consider need for long-term therapy $\begin{cases} \text{recurrent acute attacks} \\ \text{renal stones or kidney disease} \\ \text{tophi} \\ \text{severe hyperuricaemia?} \end{cases}$
    (iii) Use allopurinol for overproducers, or if renal stones or renal failure.
    (iv) Consider uricosuric drugs in undersecretors – with maintenance of good renal flow.
        Uricosuric drugs: probenecid
                             ethbenecid
                             sulphinpyrazone
                             azapropazone
                             (Calcitonin, phenylbutazone, high dose aspirin, benzofurin compounds)
    (v) Cover first three months of therapy with colchicine 0.5 mg b.d. or a non-steroidal anti-inflammatory drug.

---

subside. A number of other newer anti-inflammatory drugs appear to be effective. Phenylbutazone is widely used, but because of the tendency to cause fluid retention, and the small but significant risk of marrow suppression, this drug should not be recommended. Azapropazone in doses of 100–2400 mg day^{-1} is also effective, and has the advantage of being hypouricaemic and a potential long-term therapeutic agent as well (see below). Occasional cases of gout are slow to resolve and difficult to treat. If monoarticular, aspiration of the fluid from the joint, and injection of a small quantity of steroid may turn off the inflammatory response. In exceptional cases, or in prolonged polyarticular attacks, a short, sharp course of systemic steroids given as oral prednisolone or by ACTH injections, may be needed to terminate the attack.

Patients with gout should have an anti-inflammatory drug which they know suits them available in their pocket at all times, as therapy is most effective if started at the very beginning of the attack. Effective treatment should be able to produce a marked improvement in the symptoms within a few hours of starting the drug.

11.3.2   Pseudogout

The principles in the treatment of pseudogout are similar to those of gout. The joint should be rested, and the diagnosis confirmed by identification of crystals in the synovial fluid. The act of aspirating the joint may be therapeutic, as well as diagnostic; and, provided that there is no evidence of infection, can be combined with the introduction of a small quantity of a long-acting local steroid preparation. The response of pseudogout to colchicine is less predictable than that of gout. Anti-inflammatory drugs, such as indomethacin, and the newer non-steroidal agents, given in large doses initially, the dose being tailed off as the attack subsides, appear to be helpful in reducing the severity and duration of pseudogout. Although the condition is usually self-limiting, it may take weeks rather than days to subside. However, pain and inflammation are often less severe than in gout, and it is rarely necessary or wise to resort to other therapy, such as systemic steroids.

11.3.3   Acute calcific periarthritis

This relatively rare form of acute crystal-induced synovitis can be exceedingly painful and dramatic in its presentation. No data are available on the response to colchicine, but anti-inflammatory drugs are probably helpful. Aspiration of the inflammatory exudate has been recommended, and does appear to benefit some cases. The commonest site is the rotator cuff area of the shoulder, and in some cases it is possible to aspirate a thick toothpaste-like material, containing large quantities of the hydroxyapatite crystals. Some authors recommend surgical removal of this deposit. However, as this condition is also self-limiting we would recommend simple aspiration, combined with anti-inflammatory drugs given in high doses initially, and resting the joint until the inflammation subsides. Injection of a long-acting steroid preparation into the lesion is also helpful.

11.3.4   Chronic gout

In some cases of gout, the cause is obvious and the treatment consists simply of removing that cause. For example, the unnecessary use of a diuretic is, in our experience, a common cause of gout. Similarly, severe obesity may be the problem, and weight reduction may remove the gouty tendency as well as improving general health. However, the majority of patients have no simple obvious cause for their hyperuricaemia and gout, and decisions have to be made about their suitability for long-term therapy and the choice of a suitable agent (Table 11.1).

Gout can be treated by reducing the level of serum uric acid. This can be achieved by dietary means, by inhibiting xanthine oxidase or by uricosuric drugs. However, drug treatment means a life-long commitment for both the patient and his physician, who should monitor treatment annually. The recommended indications for this life-long treatment vary, but should probably include recurrent attacks of acute gout, renal disease or renal calculi associated with hyperuricaemia, the formation of tophi, or persistent levels of uric acid above 10 mg/100 ml. Dietary advice should include maintaining normal weight, and avoiding excesses of alcohol or high purine foods. However, treatment by strict diet is neither desirable nor very effective. Inhibition of the enzyme xanthine oxidase can be achieved with the drug allopurinol, which acts as a competitive inhibitor. The dose needed to maintain the uric acid below the saturation level varies in different patients, and will need adjustment. The usual dose is between 300 and 400 mg day^{-1}. The drug is fairly safe and free of side effects, although occasional cases of skin rashes, abdominal discomfort and blood dyscrasias have been reported. Absolute indications for the use of allopurinol in preference to uricosuric drugs include extensive tophaceous gout, gross overproducers of uric acid, and the presence of renal calculi or severe renal disease. In the absence of these indications, uricosuric drugs are an alternative to allopurinol.

There are a large number of agents which increase the renal excretion of uric acid (Table 11.1). The most widely used agents are probenecid or ethbenecid in a dose of 0.5–1 g twice a day. Sulphinpyrazone in a dose of 100 mg three or four times a day is an alternative. Of the other drugs mentioned in the table, phenylbutazone and aspirin should be avoided, but azapropazone has recently been recommended in the long-term treatment of gout, and preliminary data suggest that its uricosuric effect produces equivalent reductions in serum uric acid to allopurinol. It has the advantage of combining this hypouricaemic effect with anti-inflammatory and analgesic actions which help to relieve joint pain and inhibit attacks of gout. When a uricosuric drug is given, the patient should be advised to maintain a good flow of urine to help to avoid crystalluria. It may be helpful or necessary to alkalinize the urine by giving concommitant sodium bicarbonate or acetazolamide. Uricosuric drugs are recommended for very low secretors of uric acid, but many physicians find allopurinol preferable for the long-term treatment of most cases of gout.

The initiation of hypouricaemic therapy is accompanied by an increased tendency for acute attacks of gout, particularly in the first weeks and months of treatment. This can be prevented by the use of an anti-inflammatory, or of low doses of colchicine, such as 0.5 mg twice a day. Colchicine is particularly useful in this context, as the dose needed is below that which normally causes diarrhoea. Its mode of action is not at all clear, although it has been suggested that this prophylactic effect on acute gout may be unrelated to its anti-inflammatory action, and due to an alteration in proteoglycans. Prophylactic cover should be continued for about three months after starting hypouricaemic therapy, but is unnecessary if azapropazone is used. Patients with gout should be seen and assessed regularly, and in addition to checking the control of the uric acid level, a check needs to be kept on renal function and the cardiovascular system (see Appendix 2).

11.3.5   Chronic pyrophosphate arthropathy

This is a difficult condition to treat. As explained, no drugs are available which will prevent deposition or remove the crystals. Acute attacks can be dealt with relatively easily, but the chronic destructive joint disease is hard to manage.

Non-steroidal analgesic anti-inflammatory drugs often produce symptomatic relief to chronically damaged joints. Physiotherapy to maintain movement and muscle power may be useful, and if a single large weight-bearing joint such as the hip becomes severely damaged, replacement surgery may be indicated. All too often, however, the only treatment that can be offered is a modification of the environment to reduce the disability produced by the joint damage.

A number of alternative treatments are at present being evaluated. These include oral or intra-articular use of magnesium, and synovectomy induced by radioactive colloids. The rationale behind synovectomy is that the products of the synovial inflammation and other factors from synovial sites may be necessary to allow cartilage destruction to progress. In a recent controlled trial, we found that injection of radio-active yttrium caused a significant symptomatic improvement in knee joints in chronic pyrophosphate arthropathy. Further studies are underway, but these encouraging preliminary data suggest that removal of the synovium may be helpful; this would be in keeping with synovial enzyme release having a central role in the pathogenesis as well as the symptoms of these chronic destructive arthropathies (see Chapter 8). More controlled trial data are needed to show whether medical or surgical synovectomy will indeed prevent or slow the rate of progression of joint damage.

Joint lavage has also been recommended, and can produce a temporary improvement.

11.3.6   Calcific deposits

Deposits of hydroxyapatite can accumulate in the joints, periarticular tissues and soft tissues in a number of diseases (Chapter 8). The only proven effective therapy available is diphosphonates, usually given in the form of EHDP (ethane hydroxyl diphosphonate). They may be effective in preventing widespread calcification if given early in the development of deposits, but are generally disappointing in the treatment of hydroxyapatite deposition. Alternative crystal poisons are being explored, including the use of low dose heparin, which is known to cause osteopenia if given for a long time. A number of other drugs have been tried, including probenecid and thiazide diuretics, without producing a great deal of improvement. Surgical removal is occasionally useful, although the interference with small deposits can result in the rapid accumulation of a larger aggregate of crystals. Anti-inflammatory drugs may be useful in reducing the pain and inflammation around the periphery of a deposit, but as yet no other effective treatment is available.

11.3.7 Osteoarthritis

The treatment of osteoarthritis is symptomatic or prosthetic, i.e. the available therapy can only relieve symptoms or replace damaged joints with artificial ones.

A possible role for crystals and other joint particles in the pathogenesis of osteoarthritis was discussed in Chapters 8 and 9. One way of investigating this would be to treat the crystal deposition or its consequences, and see if osteoarthritis benefitted. Thus the possible treatments of calcific deposits discussed above could be applicable to osteoarthritis; trials of diphosphonates have already been reported, although the results were disappointing. By analogy with pyrophosphate arthropathy, and in view of the work on synovial release of proteases, synovectomy is also being investigated. Relatively early radio-colloid induced synovectomies might be worthwhile in cases of osteoarthritis associated with apatite crystals (see Chapter 8).

For more detailed coverage of the medical and surgical treatment of osteoarthritis the reader is referred to general rheumatology textbooks.

11.4 Summary and conclusions

Particle-induced acute inflammation of the joints or periarticular tissues is self-limiting and easy to treat. Chronic problems associated with crystal deposition present a greater problem, and as yet gout is the only disease which can be effectively treated. However, a number of other theoretical possibilities for treating a wide spectrum of crystal-induced diseases exist, and have been outlined. Future research is likely to result in new and effective therapy for many of the conditions described in this book. This chapter has outlined some of the theoretical background to therapy rather than detailing the full management of rheumatic diseases.

Further reading

BENNETT, R.M., LEHR, J.R. and McCARTY, D.J. (1976) Crystal shedding and acute pseudogout: an hypothesis based on therapeutic failure. *Arth. Rheum.* **19**, 93.

DIEPPE, P.A., JONES, H.E., SATHAPATAYAVONGS, B. and RING, E.F.J. (1980) Intra-articular steroids in osteoarthritis. *Rheum. Rehab.* **19**, 212.

DIEPPE, P.A., DOHERTY, M., WHICHER, J.T. and WALTERS, G. (1981) The treatment of gout with azapropazone: clinicial and experimental studies. *Europ. J. Rheum. Inflam.* **3**, 392.

DOHERTY, M. and DIEPPE, P.A. (1981) ^{90}Y injections in the treatment of chronic pyrophosphate arthropathy of the knee. *Lancet.*

FOX, I.H. (1977) Hypouricaemic agents in the treatment of gout. *Clin. Rheum. Dis.* **3**, 145.

RUSSELL, R.G.G. and SMITH, R. (1973) Diphosphonates – experimental and clinical aspects. *J. Bone Joint Surg.* **55**, 66.

SKINNER, M. and COHEN. A.S. (1969) Calcium pyrophosphate dihydrate crystal deposition disease. *Arch. Intern. Med.* **123**, 636.

WALLACE, S.L. (1974) Colchicine. *Seminars Arth. Rheum.* **3**, 369.

WALLACE, S.L. (1977) The treatment of the acute attack of gout. *Clin. Rheum. Dis.* **3**, 133.

YU, T. (1974) Milestones in the treatment of gout. *Am. J. Med.* **56**, 676.

Chapter 12

CONCLUSIONS

Problems of crystal deposition are widespread, and not confined to biology. There is such a wide solubility range for ionic compounds in water that quite small changes in pH, temperature or composition can lead to precipitation. We see this in the furring of kettles, in boiler scale, and in calcium sulphate precipitates which clog oil-wells. Given the complexity of the dissolved substances in a living organism it is surprising that body fluids do not crystallize more often.

Larger living organisms have had to develop a rigid supporting structure, a skeleton, to maintain their form. This support is provided by deposition and aggregation of crystals, mainly hydroxyapatite, on a protein framework; thus precipitation from solute is an important, beneficial process in biology. However, synovial joints, which provide movement between the rigid components of a skeleton, are more susceptible to misplaced crystal deposition than most tissues, and are easily damaged by hard, inflammatory particles. The preceeding chapters have explored these crystal deposition diseases of joints through an examination of joint structure and function, the origin of crystals in joints, the mechanisms of crystal-induced damage, and the clinical features of disease associated with the presence of crystals.

The format used to discuss these conditions is shown in Fig. 12.1. It follows the conventional view of solute excess leading to crystal deposition, and thus to disease. The advantages of this approach are that it may lead to a better understanding, and perhaps better treatment, by exploring each of the necessary steps in the chain of crystal formation leading to clinically important pathological changes. The disadvantages of the scheme lie in its brevity: it is obviously too simplistic to explain all the observed phenomena. For example, solute excess does not usually cause precipitation, and crystal deposits are often asymptomatic; even though specific diseases like gout are clearly associated with crystals. Some of the steps outlined in Fig. 12.1 are explored further in the next few sections.

255

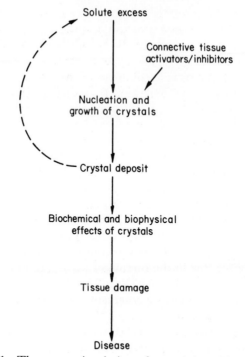

Fig. 12.1 The conventional view of crystal deposition diseases.

12.1 Solute excess: a necessary but not a sufficient cause of crystal deposition

Only a minority of people with hyperuricaemia get gout; thus solute excess is not the whole answer. Similarly, calcium carbonate deposition should occur, because the body is supersaturated for calcite, and yet it does not precipitate. Living tissues and fluids therefore appear to have a complex system of natural inhibitors of crystallization which prevent precipitates forming in most cases of solute excess.

12.2 Nucleation and growth of crystals: a sensitive process easily activated or inhibited

Crystal nucleation and growth are sensitive to fluctuations in concentration temperature, and many other local factors. In the metastable region nucleation is slow, but a short excursion into the precipitation region (see Fig. 3.12) can induce nuclei which will then grow when the system returns to a metastable situation. Similarly, since dissolution is more rapid than growth at equivalent variations from saturation, a short increase in solubility can wipe out the effects of long growth periods.

Thus short-term changes in metabolic levels, and local fluctuations, will be as important as the long-term average concentration of a solute in the body. Similarly local activators or inhibitors of nucleation and growth may have a major role to play in influencing the existence and distribution of deposits. The presence of localised solute excess may influence the precipitation of calcium pyrophosphate in cartilage, and it may be that proteoglycans and other components or connective tissue are providing natural inhibition of nucleation, helping to prevent precipitation on the surface of collagen or other fibres. In the case of hydroxyapatite, a specific system has evolved to enhance nucleation and growth; the cell derived matrix vesicles enabling local increases in solute concentration to occur, and possibly providing a surface for nucleation and growth.

12.3 The crystal deposit: a necessary but not a sufficient cause for a crystal-related disease

Precipitation of crystals is common in joints. However, many deposits, perhaps the majority, remain asymptomatic. What is it about the deposit that makes it harmful in some cases? Inflammation is the best known pathological reaction to crystals, and occurs to a varying degree with different types of crystal. To cause inflammation crystals must be free in the synovium or synovial fluid, and present in sufficient numbers, sizes and shapes to provide available surfaces to interact with cells and proteins. If the deposit remains embedded in cartilage, inflammation does not occur and other pathological reactions depending on cellular interactions will not proceed.

Crystal deposits are hard, and change the mechanical properties of tissues; they may also act as wear particles to joint surfaces. As in the case of inflammation, the degree to which damage of this sort proceeds may depend on the integrity of the cartilage and other aspects of the joint, as well as on the nature and site of the deposit. The wear debris may also cause inflammation. Crystal-induced damage therefore depends on many local factors.

12.4 Crystal-related arthropathies: self-limiting inflammation, or chronic destructive joint damage

Crystals in joints are associated with diseases which are clinically, radiologically and pathologically distinct from other joint diseases. Two main reactions occur: acute, self-limiting episodes of inflammation in one or more joints, or a chronic destructive process. Thus crystals are a marker of specific rheumatic diseases, which between them account for a considerable amount of suffering and disability.

12.5   Distribution of disease is not readily explained

Not all joints are involved (for example the temperomandibular joint is rarely affected by crystal deposits), and the peculiar distribution of these diseases remains hard to explain. Age is an important factor, most crystal-related arthropathies being more common in the elderly. This may be due partly to slow crystal growth, and partly to other connective tissue changes occurring in the elderly; whatever the causes, it means that these diseases will continue to rise in importance as the community ages.

Can the factors mentioned be accommodated into one hypothesis? Can the scheme in Fig. 12.1 be improved upon, to provide a new framework to aid understanding and research in these important diseases?

An hypothesis is presented in Fig. 12.2. It proposes that solute excess predisposes to crystal deposition within altered connective tissue, and that damaged joints are more susceptible to crystal-induced damage than normal ones.

Thus any disease affecting connective tissue might reduce normal inhibition to crystallization, or unmask activators. Ageing alone may also contribute in this way, lowering the threshold of precipitation in the presence of localized solute excess. A deposit trapped in relatively normal joint cartilage causes little damage, but if the joint is already diseased crystals

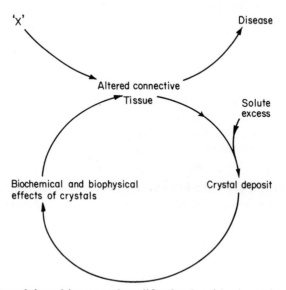

Fig. 12.2   Crystal deposition as an 'amplification loop' in the pathogenesis of the disease.

will escape more easily, and accelerate inflammatory or mechanical damage more readily.

Crystal deposition is therefore seen as an 'amplification loop' capable of accelerating, as well as initiating, joint disease. This concept does not apparently apply to inflammation caused by extrinsic crystals; in such cases the source of the precipitate is essentially external to the joint. With pyrophosphate and phosphate deposits there is a much closer link between metabolic and mechanical disturbances within the cartilage itself and the precipitation.

This allows us to look on crystals as a mechanism of joint damage, rather than regarding a crystal as synonomous with a disease. This book has aimed at identifying underlying principles to crystal-induced joint diseases. For the pyrophosphate and the phosphate diseases we have found such principles lacking; the hypothesis presented in Fig. 12.2 therefore seems justified as a basis for further investigations.

Crystal deposition is one of a limited number of ways in which connective tissues respond to an insult; particles are one of a limited number of insults which can damage joints.

Appendix 1

PRACTICAL HINTS ON USE OF THE POLARIZED LIGHT MICROSCOPE FOR CLINICAL RHEUMATOLOGISTS

The principles behind the use of the polarized light microscope to identify crystals has been outlined in Chapter 3. Further, more detailed references are appended below. This section is intended as a practical guide to help everyday usage of the microscope in rheumatological practice.

A1.1.1   Equipment

An ordinary microscopy can be converted to polarized light using pieces of polaroid sheet. However, a microscope designed for polarized light is preferable since it will allow centred rotation of the sample and the convenient insertion and removal of compensators and polarizers. The identification of small crystals requires complete extinction, total blackness, with crossed polarizers. Thus strain-free achromatic objectives and high-quality polarizing filters are preferable. The most useful range of objectives are × 16, × 40 and × 100, oil immersion. A powerful light source will greatly enhance the rate of crystal identification. Thus at least a low-voltage high-intensity microscope lamp should be used although a quartz iodine or even a Xenon arc would be preferable for critical work.

In addition, it is necessary to have an eyepiece with a crosswire and graticule to measure the size and angles of crystals; an analyser and polarizer, one of which should rotate; a first-order red compensator and a graduated rotating stage. The microscope should be kept covered and scrupulously clean, left in one place, and used by a small number of properly trained people.

A1.1.2   Other equipment

Other essential items include:
   (a) glass slides, cover slips and lens papers;
   (b) pipettes and/or wire loops;
   (c) nail varnish and slide labels;
   (d) an aerosol clean-air jet for dust removal.
Additional items which help to improve diagnostic reliability include photomicrography equipment and plenty of storage space for slides, synovial fluids, pictures and reports. A bench centrifuge, refrigerator and freezer are also useful, but not essential for diagnostic use.

A good rule (which we find difficult to keep) is to be obsessionally clean and tidy every time you use the apparatus. Dust and dirt are the biggest enemies of synovial fluid polarized light microscopy.

Before use the microscope should be adjusted and centred so that a crystal can be rotated without moving from the field of view and a control crystal remains central when the lenses are changed. The illumination should also be adjusted to a centred, uniform bright field. This is done by centering and focusing the condenser. In subsequent use the lamp is run at close to its maximum intensity and the illumination controlled by reducing the substage (condenser) iris diaphragm. The lamp iris should be closed down as far as possible without entering the field of view; this reduces haze due to scattering from parts of the sample outside the field. More details on microscope adjustments can be found in Hartshorne and Stuart (1970) (see Chapter 4).

A1.1.3   Technique

Joint aspiration is carried out in the usual way. In general, 10 ml of fluid is desirable, but any amount, even the contents of a needle from a 'dry tap' will do. Where plenty is available, immediately divide it up as follows:
   (a) put 4 ml into an EDTA container for a total and differential white cell count. Total counts can be done by a 'Coulter Counter'. The differential count is very useful, and should always be measured;
   (b) put 5–10 ml in a sterile container and send for bacteriological examination. Crystal-induced synovitis and infection can co-exist, and fluid culture is mandatory in all cases;
   (c) put one small drop of fresh, uncentrifuged fluid on a glass slide, cover with a cover slip and seal the edges with nail varnish to prevent the slide drying out. All glassware should be cleaned with lens paper and the air jet,

and the drop of fluid applied with a sterile wire loop, or small, disposable pipette. Examine as soon as possible; if this is not immediately, label the slide;

(d) keep the remainder of the fluid in a plain container without any anticoagulant or preservative. If it is kept overnight, it is best at 4° C.

When examining the wet slide preparation, it is probably best to follow a set sequence on every occasion. One then gets used to examining all the features of the specimen, and can note them down in a logical order. We follow this plan: first examine the slide using a × 16 objective without the polarizer or analyser. This enables a quick appraisal of the contents: the cells, cartilage fragments, clots, fibrin, etc. Further examination with the × 40 lens enables more cell details to be noted, although wet preparations, without stain, only allow limited cell morphology to be observed.

Once the general fluid characteristics have been ascertained, one can use the polarized light. It is best to examine several regions of the slide closely, looking particularly at leucocytes and any clots, as these tend to trap crystals. (If possible a fluid clot should preferentially be put on the slide.) The analyser and polarizer are then 'crossed' to give a dark background and as *much light as possible* focused on the slide. The slide should then be searched with the × 40 objective. If the background is completely dark, the compensator can be slipped in and out to help check that one is maintaining the focus. The fine focus should be continuously racked slightly up and down, as this often helps to bring otherwise 'invisible' birefringent particles into sight.

When a birefringent particle is seen, shining out as light against the dark background, focus on it, and then insert the first-order red compensator. Focus up and down through it, and try to define whether it has crystalline symmetry. Beware of amorphous dust particles, or strands of multi-coloured dusty contaminants. Crystals usually have sharp, definable edges, and straight sides; they are not ragged or irregular in outline.

If crystals *are* seen, one should note the following features.

(a) General: overall numbers (lots, a few, one, etc.), distribution (in clots only, extracellular, intracellular), uniformity (lots of different ones, they all look the same, etc.).

(b) Particular: look at two or three representative crystals, and examine them closely to record:

    size,
    shape,
    (extinction angle),
    pleomorphism (shape variability),
    degree of birefringence (strong or weak),
    sign of birefringence.

Size and shape are recorded with the aid of a graduated stage and eyepiece graticule. You may need the oil immersion lens to look at a crystal's angles to define whether a crystal is, for example, monoclinic (one right angle) or triclinic (unequal angles and sides).

The extinction angle is that at which the crystal becomes invisible on rotation of the stage (Fig. 4.8(b)). It is not essential for identification, but can help to define a particle as a purely crystalline specimen.

When considering the degree of birefringence remember that bigger crystals are always brighter. Beware the very small, almost isotropic crystal (i.e. almost no bright light on dark background).

The sign of birefringence is important, and is assessed as follows (Fig. 4.10).

(1) Check that the long axis of the crystal is an extinction direction, that the crystal becomes dark between crossed polars when the long axis is parallel to one of the polars (not always true, and not always easy to define the long axis!).

(2) Align the long axis with the slow vibration direction of the compensator (marked ↗ on the handle).

(3) Note whether the colour change is to blue or to yellow from the pink colour given by the compensator alone.

(4) Try and remember which colour means which sign! Blue is positive and yellow is negative.

It is useful to record all of this information on a report form (one is shown below), and interpret the findings separately, rather than writing 'gout crystals seen' or other inaccurate conclusions.

The fluid kept may be needed for a second slide preparation to check on other features. Remember, however, that although crystals usually survive in fluids for several days or weeks; fluid kept on the bench can grow crystals that were never there *in vivo*. Early examination is therefore desirable. It is rarely necessary to centrifuge fluids, although examination of a clot is often useful.

If there is a particularly strong suspicion that crystals are present, but none can be seen, the following manoeuvres may help to visualize them:

(1) use the oil immersion lens, some fluids have very tiny crystals (it has recently been shown, by the use of electron microscopy, that the occasional gouty synovial fluid, with no crystals visible by polarized light microscopy contains crystals below the size of optical resolution);

(2) rack the condenser down, or use a phase-contrast objective lens to produce partial phase effects; this sometimes helps to visualize the crystals;

(3) re-examine clots and white cells;

(4) examine a centrifuged deposit from the fluid.

Another technique that may come in useful is to prepare a smear, in a similar fashion to a blood film, and stain with Wright's stain or a similar preparation. Crystals may get dissolved or washed off these slides during staining, but the preparation can be kept indefinitely, and allows cell morphology and differential white counts to be examined.

A1.2   Particles found in synovial fluids (Table A1.1)

A1.2.1   Monosodium urate monohydrate

Typically 2–10 μm long, needle-shaped, strongly negatively birefringent crystals. Often numerous, with many intracellular crystals. Can be dissolved with uricase.

A1.2.2   Calcium pyrophosphate dihydrate

Very variable in all respects. Size ranges from less than 1 μm, up to about 10 μm long. Most are either monoclinic or triclinic (acicular or rhomboid). May be lots, but often only one or two can be seen on any one slide; often seen in fibrin clots. May be very weakly birefringent. Most have positive sign of birefringence. Often appear to have a chip out of one corner, due to 'twinning'.

A1.2.3   Cholesterol

Thin, adherent plates, may be coloured. Usually 5–10 μm long, regular and square. May appear to have a chip out of the corner. Often very numerous when present and may make the fluid look like milk. Weak birefringence. Can be dissolved with lipid solvents such as alcohol, added in tiny amounts to the slide.

A1.2.4   Steroids

Vary with the type of intra-articular steroid used. Usually appear as many tiny, strongly birefringent particles. May look like pyrophosphate, but usually much more optically active.

A1.2.5   Dicalcium phosphate dihydrate

Occasionally seen, usually in fluids left on the bench before examination. Large (10–20 μm) strongly positively birefringent crystals which may have pointed ends, and usually appear in star-shaped clusters.

A1.2.6   Hydroxyapatite

Usually indistinguishable from 'dust' in the light microscope. May appear

Table A1.1

Crystal	System	Birefringence	Main IR peaks (600–1400 cm^{-1})	X-ray Spacings* d(Å)	Intensity I
Monosodium urate Monohydrate $NaH.C_5H_2O_3N_4.H_2O$	Triclinic	Negative	1380 1350 1260 1010 800 770 745 725 600	9.29 7.75 3.18 3.12	30 80 100 90
Calcium pyrophosphate Dihydrate $Ca_2P_2O_7.2H_2O$	Triclinic	Positive	1160 1135 1026 1005 936	8.01 6.95 3.21 3.10	vs s ms ms
	Monoclinic	Positive	958 580	7.37 4.62 3.22 3.03	s s s s
Hydroxyapatite $Ca_{10}(PO_4)_6(OH)_2$	Hexagonal		1092 1065 1028 633 603	8.17 2.81 2.78 2.72	11 100 60 60
Brushite $Ca\,HPO_4.2H_2O$	Monoclinic	Positive	1215 1133 1057 870 580	7.62 3.80 3.06	100 30 8
Octacalcium phosphate $Ca_8\,[H_2(PO_4)_6].\,5H_2O$			∼1020 600 ∼560		
Cholesterol Cholest–5–en–3β–ol $C_{27}H_{46}O$			1376 1052 1020 955 837 797	33.6 5.74 5.09 4.90	100 21 20 21

*First and three strongest lines.

as tiny birefringent clumps. Sometimes visualized intra-cellularly with a surrounding 'halo' in stains, smears or fluids. Will stain with calcium stains such as Alizarin red.

A1.2.7   Other particles

Cartilage fragments may appear to have slight birefringence, but have rough edges. Fibrils may also cause confusion. Birefringent amyloid fragments have been identified occasionally. Metal and plastic fragments are sometimes seen in fluids from prosthetic joints. Calcium oxalate and lithium heparin crystals can occur if these anticoagulants are used. Thorns have been identified in plant thorn synovitis. Finally, fragments of glasswear or other laboratory contaminants may occur, and dust particles of every size, shape and description are seen.

    The beginner is advised to obtain urate crystals from a tophus, and pyrophosphate crystals from a post-mortem piece of cartilage so as to get used to the optical properties of these all-important crystals in his own microscope.

A1.3   Reporting the findings

Do not write 'pyrophosphate crystals seen', you might be wrong! It is better to describe the findings, and add an interpretation. A specimen reporting form is reproduced opposite. Although at first sight long, the use of boxes to tick makes it quick and easy to fill in and any other available investigations can either be recorded as having been done, for future reference, or filled in if available.

Reporting form for synovial fluid analysis
Patient identification:
Name.. Age..

Hospital No. .... Sex ....

Consultant ....

Joint aspirated: Date: Time:
Reason for aspiration:
Clinical diagnosis:

Report to be sent to:

Quantity of fluid received:...................ml
Appearance:
Viscosity: High ☐   Intermediate ☐   Low ☐
Fluid stored: Yes ☐   No ☐
Stained smear preparation: Yes ☐   No ☐
Wet film:
      General appearance:
      Cells:
      Clots: Numerous ☐   Few ☐   None ☐
      Cartilage fragments: Numerous ☐   Few ☐   None ☐
      Fibrin: Numerous ☐   Few ☐   None ☐
Birefringent particles: Present ☐   Absent ☐
              Numerous ☐   Some ☐   Few ☐
              Location: Extracellular ☐
                       Intracellular ☐
                       Clots   ☐
                       Other   ☐
Typical characteristics:
  Size:
  Shape:
  Extinction angle:
  Anisotropism: Strong ☐   Intermediate ☐   Weak ☐
  Signs of birefringence: Positive ☐   Negative ☐

Other investigations	Requested	Result
Total white cell count	☐	....
Differential	☐	....
Bacteriology	☐	....
Electron microscopy	☐	....
Rheumatoid factor	☐	....
Cell morphology	☐	....
Others	☐	....

Conclusions:

Name ....   Signature ....

INVESTIGATING PATIENTS WITH GOUT OR HYPERURICAEMIA

If hyperuricaemia or gout are suspected, three questions must be answered.
(1) Does the patient have true hyperuricaemia?
(2) What is the cause of the hyperuricaemia?
(3) Is the hyperuricaemia causing any damage?

A2.1.1   Does the patient have true hyperuricaemia?

As explained in Chapter 6, a serum urate level of 7 mg/100 ml or more represents supersaturation in physicochemical terms. Several epidemiological studies have been carried out, and they suggest that about 5% of the adult male population of Europe or the United States will have hyperuricaemia if it is defined in these terms.

Racial, temporal and seasonal variations occur. The Maoris of New Zealand and some other Polynesians have higher levels than most races, as do Australian Aboriginals. Repeated estimations in the same individual may show occasional high recordings (perhaps related to episodic dietary excess), and in one study a seasonal variation was recorded, serum urate levels being highest in the summer.

Urate levels vary in different laboratories. Chemical methods and autoanalysers usually depend on the reduction of tungstate compounds by uric acid, and give levels that probably overestimate true serum urate by about 0.5 mg/100 ml. Enzymatic methods depend on uricase breaking the urate down to allantoin, and give rather lower levels. Large comparisons have been carried out, and show a good correlation between the methods; a coefficient of variation of about 5% is usual. It is therefore important to be aware of the methods used, variance encountered, and any tendency to over- or underestimate true urate levels in the local laboratory.

A diagnosis of true hyperuricaemia can only be made if two or more separate serum urate estimations show a level above the saturation of uric acid in serum, this will be about 7.5 mg/100 ml in most laboratories using autoanalysers.

A2.1.2   What is the cause of the hyperuricaemia?

Hyperuricaemia is due to increased intake, overproduction, reduced excretion, or a combination of these factors.

(a) Increased intake?

A diet very rich in meat, meat extracts or offal can raise the serum uric acid. Excess alcohol can also cause hyperuricaemia, although this is probably due to hyperlacticacidaemia suppressing excretion, to the ethanol increasing production, or occasionally to the lead in the drink (as in old port), and not because of the purine load in the beverage. Excess calories cause obesity, and this in turn results in hyperuricaemia, although the mechanism is not clear.

(b) Overproducer or undersecretor?

It is important to classify hyperuricaemic individuals into these two categories. This is best done by estimating the total excretion of uric acid while on a controlled diet with a known purine content. In practice it can be achieved by using the following regime: institute a low-purine, alcohol-free diet for one week; collect two 24 h urine samples on the last two days of the diet, and estimate the total uric acid content. Normal excretion varies from 300 to 600 mg/24 h; overproducers will excrete more than 600 mg both days, undersecretors less than 300 mg (Table A2.1).

A simpler, much easier, but less exact way of assessing uric acid excretion uses a single, mid-morning sample of urine and serum. Urinary uric acid is multiplied by plasma creatinine and divided by urinary creatinine, to provide an estimate of the uric acid excretion/100 ml glomerular filtrate. Normal levels vary from about 0.3 to 0.5 mg/100 ml, and the method identifies most overproducers. The uric acid/creatinine ratio on spot urine samples can also be measured, and may help (it should be less than 0.5).

Urine estimations of uric acid are beset by many problems. Significant errors can be caused by ill-timed or incomplete collections, by bacterial contamination or by crystalluria. Estimations will be inaccurate if the creatinine clearance is less than 40 ml min^{-1}, or if the urinary flow is less than 1 ml min^{-1}. It is therefore important to exclude infections, take great care with the collections, to know the urinary flow and renal function, and avoid discarding sediment, which may contain large quantities of crystalline

Table A2.1 Investigating hyperuricaemia and gout

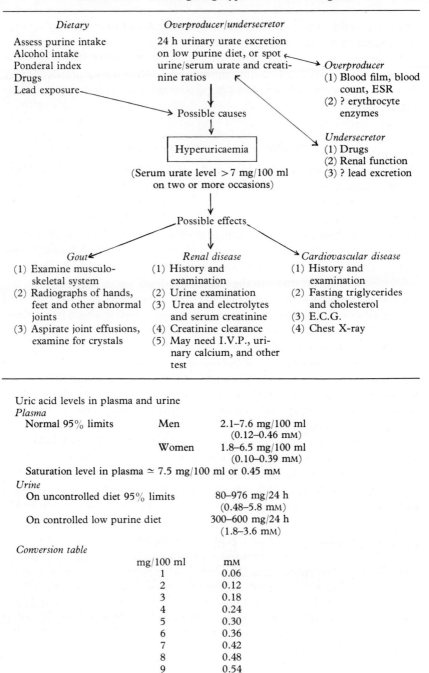

Dietary
Assess purine intake
Alcohol intake
Ponderal index
Drugs
Lead exposure

Overproducer/undersecretor
24 h urinary urate excretion
on low purine diet, or spot
urine/serum urate and creatinine ratios

→ Possible causes

Hyperuricaemia

(Serum urate level >7 mg/100 ml
on two or more occasions)

Overproducer
(1) Blood film, blood
    count, ESR
(2) ? erythrocyte
    enzymes

Undersecretor
(1) Drugs
(2) Renal function
(3) ? lead excretion

Possible effects

Gout
(1) Examine musculo-
    skeletal system
(2) Radiographs of hands,
    feet and other abnormal
    joints
(3) Aspirate joint effusions,
    examine for crystals

Renal disease
(1) History and
    examination
(2) Urine examination
(3) Urea and electrolytes
    and serum creatinine
(4) Creatinine clearance
(5) May need I.V.P., uri-
    nary calcium, and other
    test

Cardiovascular disease
(1) History and
    examination
(2) Fasting triglycerides
    and cholesterol
(3) E.C.G.
(4) Chest X-ray

Uric acid levels in plasma and urine
Plasma

Normal 95% limits	Men	2.1–7.6 mg/100 ml (0.12–0.46 mM)
	Women	1.8–6.5 mg/100 ml (0.10–0.39 mM)

Saturation level in plasma ≃ 7.5 mg/100 ml or 0.45 mM

Urine

On uncontrolled diet 95% limits	80–976 mg/24 h (0.48–5.8 mM)
On controlled low purine diet	300–600 mg/24 h (1.8–3.6 mM)

Conversion table

mg/100 ml	mM
1	0.06
2	0.12
3	0.18
4	0.24
5	0.30
6	0.36
7	0.42
8	0.48
9	0.54
10	0.60
12	0.72
15	0.90
20	1.20

urate. As with the serum estimations, at least two collections are recommended.

As shown in the table, the causes to look for in overproducers and undersecretors are quite different. Overproduction may be related to obesity, glycogen storage diseases, or psoriasis; but the two main possibilities that need exclusion are specific enzyme defects (especially in younger patients) and neoplastic disease (especially myeloproliferative diseases in the older age group). A full history and clinical examination are obviously necessary, and a blood count, film examination and ESR should always be performed. If the patient is unusually young, has a strong family history, or where there is no other obvious cause for overproduction, enzyme defects should be suspected. Some specialist centres have methods available to assay HGRPT and other enzymes in erythrocytes.

If urinary urate is normal or low, possible causes of the hyperuricaemia include drugs, lead, and renal disease. All patients should have their urea and electrolytes, and serum creatinine, measured. The urine can be examined for the presence of proteinuria or casts; examination for crystals is more difficult. If the urine is allowed to cool uric acid crystals often form, and many patients produce a 'sludge' containing calcium phosphates or oxalate. A realistic search for urinary urate crystals requires collection in a thermos flask, and examination of urine on a microscope stage heated to 37° C. If lead poisoning is suspected (as in 'moonshine' drinkers), measurement of erythrocyte α amino-levulinic acid dihydratase activity may help (reduced in lead poisoning). Alternatively, the urinary excretion of lead after a dose of calcium disodium ethylenediaminate can be assessed; values in excess of 500 μg/24 h indicate increased tissue accumulation of lead.

In large series of gouty patients, most authors have found a small minority of overproducers, and less than 5% have one of the described enzyme defects; similarly only a small number have primary renal disease, lead poisoning or another primary disease. The majority have normal or reduced urinary urate output, with no obvious cause, or they are on diuretic therapy.

A2.1.3 Damage caused by the hyperuricaemia

The chief disease associations of hyperuricaemia are gout, renal disease or urinary calculi, cardiovascular problems, and tophi.

The musculoskeletal examination will reveal any chronic joint damage, and a search for tophi should include the ears, elbows, achilles tendon and other common sites. Radiographs of the hands and feet and any other abnormal joints can be examined for evidence of 'punched out' gouty erosions or other joint diseases. A recent report suggested that aspiration of asymptomatic first metatarsophalangeal joints may reveal crystals, although this has not been our experience. If acute arthritis occurs joints should certainly be aspirated and examined for evidence of monosodium urate monohydrate crystals. The prevalence of hyperuricaemia is far higher than the prevalence of gout, so joint pain in hyperuricaemic patients often has another cause ('non-gout'), and accurate diagnosis is important.

If there is evidence of renal impairment, or a history of ureteric colic or stone formation, an intravenous pyelogram should be performed. Uric acid stones are radiotranslucent, but radio-opaque calcium urolithiasis is also common in hyperuricaemia (Chapter 10), and urinary calcium excretion may need to be measured. Hypertension, obesity, hypertriglycridaemia (usually type IV), and cardiovascular disease all have an increased frequency in hyperuricaemia, although their relationship to each other is not clear. Fasting blood lipid levels, an E.C.G. and chest X-ray should be performed in addition to the other tests mentioned. An annual review of patients with hyperuricaemia needs to include an assessment of cardiovascular and renal function.

Appendix 3

INVESTIGATING PATIENTS WITH CALCIUM DEPOSITION IN JOINTS

Three different situations may result in the suspicion that a patient has calcium-containing crystals in his joints. (1) First, a clinical presentation typical of pseudogout, acute calcific periarthritis or chronic pyrophosphate arthropathy may warrant a search for crystals; (2) secondly, radiographic changes may include linear or spotty deposits of calcific density in or around the joints, suggesting pyrophosphate or hydroxyapatite deposition; (3) finally, suspicion may be raised by seeing abnormal cartilage or synovial deposits at arthroscopy or arthrotomy, or from finding crystals on examination of synovial fluids and biopsy material under polarized light (Table A3.1).

When calcium deposits are suspected, three questions must be answered.

(1) Are any crystal deposits present; if so, what is the likely species of crystal, and what is the distribution of the deposits?

(2) Is there an obvious cause for the deposition; is it due to local tissue damage or to a generalized metabolic disorder, and is there a reversible element to the cause?

(3) What effects are the deposits having; are they causing acute inflammation, or contributing to chronic joint damage?

A3.1.1 The nature of the deposition

The presence, type, and distribution of any crystalline deposits can be investigated by joint radiography, synovial fluid analysis, arthroscopy or biopsy. A knowledge of the likely salts, and their radiographic features and distribution is helpful (Fig. A3.1).

273

Table A3.1 Investigating the presence of calcium-salt deposition in joints

---

(a) Reasons for suspecting the diagnosis

    (1) Clinical presentation: e.g. acute arthritis of the knee in an elderly person is likely to be pseudogout.

    (2) Radiographic appearances: Linear deposits in menisci or hyaline cartilage, or 'spotty' articular and periarticular shadows of calcific density.

    (3) Pathology: Deposits seen at arthroscopy or arthrotomy

                                   or

                      Crystal deposition seen in histological sections

                                   or

                      Crystals seen under polarized light or synovial fluid.

(b) Questions to be asked about articular deposits

    (1) The nature and distribution of deposits

    (2) The cause: local or general metabolic abnormality? reversible or irreversible?

    (3) The effects: acute crystal-induced inflammation? Are deposits contributing to tissue damage?

Radiographs should include a single view of the hand and wrist, an AP and lateral of the knees, and a pelvic X-ray to include the hips if pyrophosphate deposition is suspected. The typical linear deposits are most common in the knee, and grow preferentially in fibrocartilage pads such as the knee menisci, triangular ligament of the wrist and pubic symphisis. Periarticular hydroxyapatite deposits are commonest around the shoulders and hips, but can occur almost anywhere.

Suspicious joint effusions should be aspirated and examined under polarized light as described (Appendix 1). Deposits may also be seen at arthroscopy or arthrotomy and biopsied for examination using standard histological techniques, with or without the help of polarized light or analytical electron microscopy (Chapter 4).

A3.1.2 The cause of the deposition

Deposits may be ideopathic, age-associated, familial, or due to local tissue damage or a generalized metabolic abnormality (Table A3.2). As the table shows, the causes are different in the case of pyrophosphate or hydroxyapatite deposits.

A full medical history and examination should be done, with emphasis on any joint diseases and joint laxity, drugs and metabolic abnormalities.

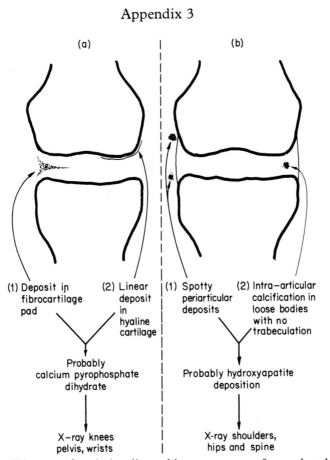

Fig. A3.1 Diagram of typical radiographic appearances of pyrophosphate and hydroxyapatite deposition.

Useful screening tests include serum calcium and thyroid function tests. The urea and electrolytes and serum creatinine will exclude renal disease, and other metabolic causes of deposition such as haemachromatosis, hypomagnesaemia and hypophosphatasia, are very rare, and usually clinically obvious when present. (In a series of 105 patients, we only found the calcium and thyroid function tests of any help in detecting unsuspected abnormalities.)

A3.1.3   Are the deposits causing any damage?

Many calcific deposits remain asymptomatic for years, if not for life. The two chief disease consequences of articular crystals are acute inflammation, and chronic joint damage.

Table A3.2 Causes of articular calcific deposits

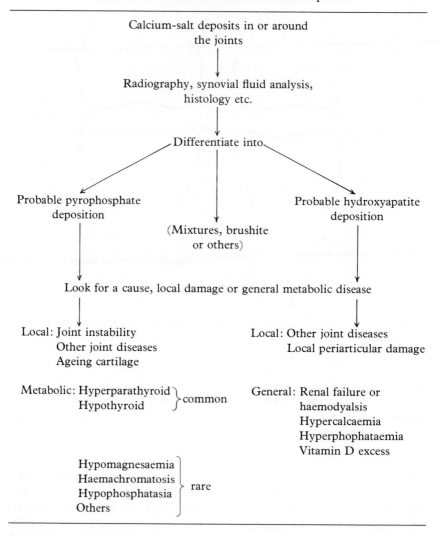

(a) *Acute inflammation*

Diagnosing acute inflammation presents no problem. The difficulty lies in knowing whether or not it is due to crystals. Calcific periarthritis is rare, can occur at any age and any site, although the shoulder predominates. Pseudogout is principally a disease of elderly people, and usually involves the knee, wrist or ankle, although younger men are occasionally affected. However, radiological chondrocalcinosis is present in about 30% of all people over the age of 70, and will obviously sometimes be present in joints affected by other causes of inflammation.

Other important causes of acute inflammation in the elderly include gout (especially if the patient is on diuretics), sepsis, acute rheumatoid disease of the elderly, and neoplastic diseases. Further confusion arises in old people as various inflammatory conditions, including pseudogout, can present with polymyalgic symptoms. In younger people the differential diagnosis is easier, and involves different diseases such as Reiter's syndrome, gonococcal arthritis and sero-negative spond-arthritides.

Joint radiography is helpful, and a serum uric acid will alter the degree of suspicion about gout. However, the only definitive investigation in acute arthritis is examination of synovial fluid as outlined in Chapter 4 and Appendix 1. Gout, pseudogout or sepsis can all coexist, so finding one type of crystal does not exclude other diseases. Acute arthritis is best diagnosed on the basis of its pattern, and the patient's circumstances (age, injury, etc.) and from synovial fluid analysis.

(b) *Chronic joint damage*
It is almost impossible to say whether joint damage in any particular patient is being caused by crystals. Reasons for believing that deposits will contribute to chronic destructive changes have been outlined earlier in this book; the radiographic appearances, distribution of changes (e.g. bad osteoarthritis of wrists or shoulders), or progression of the disease (i.e. rapidly destructive or particularly severe osteoarthritis) may all suggest that calcium salts are present, and aggravating the situation. This question will become more important if treatment capable of removing the deposits were to become available.

INDEX

278